THE WATER CARRIERS

GREGORY GREUNKE

Greunke Haus

This is a work of fiction. Any resemblance to actual persons, living or dead, events, or locations is purely coincidental.

Greunke Haus
www.greunke.com

Book cover by Tim Barber

Library of Congress Control Number: 2024925793
ISBN: 979-8-9918682-2-8 (ebook)
ISBN: 979-8-9918682-1-1 (hardcover)
ISBN: 979-8-9918682-0-4 (paperback)
ISBN: 979-8-9918682-3-5 (audiobook)

Version 1.2

To my wife, Tammy. I couldn't have written this story without your laughter and love.

To my mom, who taught me to love reading, and to my dad, who taught me to love water.

Ronan and Isla, I wrote this story so that you don't inherit a world like this...although the AugWatch would be pretty cool.

Contents

Chapter One

A CAR SHOT across the causeway, and the charcoal forest shimmered. The desiccated remnant of something that once slithered, now barely distinct from the asphalt, flipped up behind the wheels as they turned. The car swerved to the side and rolled down the bank, the rhythmic grinding of the gravel slowing until it came to a stop.

In its smooth voice, Ryan, Naomi's AI companion, said, "Hey, Naomi. I have to turn the car off for a sec."

Inside the car, Naomi and Bai sat across from each other. Their eyes wide open.

"Shit! Ryan, why did you turn the car off?" Naomi asked.

"The battery cells are overheating and there's a tiny chance there could be a thermal activity."

"What does that mean?"

"Basically, the batteries could catch fire. The batteries are located below the seats and a fire under the car would be very dangerous."

The girls stared at each other. Bai wiped a drop of sweat that had already formed on her temple. "It's getting hot in here."

"Ryan, turn on the air conditioning, please," Naomi asked.

"I'm sorry, but the air conditioning won't work right now, Naomi."

Naomi sat up. "Won't work? How are we supposed to sit here in this roasting car?"

"If I were you, I'd find an area in the shade."

The virtual scene had turned off, and the girls looked out the window. Tangled trees stuck out of the black earth.

"These trees are all dead and torched from the fires," Naomi said. "There's no shade for miles."

"Let's walk back to the overpass," Bai suggested.

"Are you kidding? It'll take an hour to walk back. I'm not doing it."

"Fine. We'll just wait then."

The girls sat in silence. Their chests began to feel warmer, and they could feel the heat spreading to their belly. The air was dry and stale as the heat baked the air inside the cabin. The sides of their faces warmed.

Naomi stared, mesmerized by the sweat forming above Bai's lip. She sagged a little, rested her hand on the seat beside her, and yelled, "Shit, the seat is already burning hot!"

Bai barely registered Naomi's outburst. Her body was sending fluids to the surface of her skin to cool down, depleting her energy. Her eyes drooped, and she blinked, fighting back sleep.

"Too hot to sleep. Too hot to move," Bai mumbled. She stared up at the ceiling, her mouth open, taking long, slow breaths. "Open the window, Naomi."

Naomi leaned over and touched the control panel on the door. The window slid down, and hot, humid air rushed in. For just a moment, the moving air felt cool, then like a sauna, making their arms and legs sticky.

Bai's eyes shot open. "Close it. You'll let all the hot air in."

"What's the difference? It's just as hot in here."

"Well, now it is. We're screwed."

Bai tapped the door, which slid up and folded in on itself. It was a clever design for tight parking lots but provided no shade. She recoiled from the hot air like she had seen a rabid dog, then jumped out into the sun. She raised her hand to shield her eyes and scanned up and down the road. A car passed, then another.

"I can't see the overpass. It must be miles behind us," Bai said.

Another car came up on the horizon. She waved her hands over her head, but the car sped by. It was impossible to know if its occupants were awake or sleeping. The inside windows were all digital—they were probably watching a beautiful forest or meadow scene or perhaps a submarine cruising through a kelp forest, which was very popular then. She felt lightheaded and turned back to the car.

Naomi slowly raised her head. Her lids dropped, and she stared through Bai. "Nobody's going to stop unless you stand in the road." She wiped the sweat off her face with her hand and let her arm drop. The seat made a soft, hollow *thunk* sound. She was reminded immediately how hot it was, cursed, and yanked her hand back. "Why are you still standing outside? It's hot out there."

"I know. I just told you it's hot."

Naomi closed her eyes and opened them again with a slight flicker. "And you have to stand in the road."

"I what?" She took two steps forward.

"I said you have to—"

Bai rested her forearm on the door. The instant her skin touched the metal, she shrieked. It burned her arm, which stuck a little as she pulled it back. The wound was a fleshy pink and shiny, like plastic.

Instinctively, she clutched her arm to her belly and cradled it. She wailed, "Oh my god!" and dropped to her knees. The rocks were sharp. The sun had raised their temperature to over one hundred and thirty degrees. She stared at her knees, confused. She jumped up. Her knees were raw; the rocks had burned through layers of skin, some still sticking to the flesh. She staggered forward, slamming her knee into the door frame. The fragile skin tore, and she scrambled into the car.

Her screams were primal. Her elbow collapsed, and she fell. She lifted herself again, flopped onto the bench seat in a tight curl, and shivered convulsively. Then she rolled onto her back with her hands held over her face. She repeatedly clenched and stretched her fingers wide. She sobbed and screamed.

Naomi rocked back and forth, mumbling, "No, no, no." She grabbed her hair and pulled, desperately trying to process what was happening. Her mind went blank, and then took her back to the charging station from an hour earlier.

After plugging the car into the charging station, the girls had run toward the convenience store. They laughed when they pushed open the door, and the cool air washed over them. Naomi skipped directly to the wall of refrigerators at the back and leaned her forehead against the glass, rolling her head back and forth. She lifted her head and glanced at Bai, who was rifling through the snack aisle with a look of distaste.

Naomi straightened up and called, "Want a pop?"

Bai studied the snacks. "You know how much sugar's in those things? I'd rather eat dirt." She grabbed an IE Gel and tossed it to Naomi. "Here, this'll keep us going."

Naomi caught it. Her hand trembled slightly. As she tied her hair back, she noticed the man behind the counter watching them intently. His baseball cap was stained with sweat and oil, and a patch reading "Fuelless" was barely legible through the grime.

"See something you like?" she snapped, more to break the tension than out of any real annoyance.

Bai was standing behind her now. She pushed her knee into the back of Naomi's leg and hissed. "Naomi, chill. We're in the middle of nowhere. He could be a psycho killer."

The man's smile didn't reach his eyes. "Just curious. Where you girls headed?"

"University of Georgia." Reflexively, Naomi pushed out her hip, pumped her arm in the air, and yelled, "Go Bulldogs . . ." Her voice trailed off as she realized how inappropriate the gesture was.

The girls placed their items on the counter. The man's eyes lingered momentarily before he looked down at the counter to grab a small wand. "Ah, right. Thanksgiving break." He waved the wand over the items. It made a sharp *beep* as it registered the cost for everything.

He looked up again and pushed the items back towards the girls. Naomi held up her AugWatch, a large, rectangular gold device fastened to her wrist with two white bands, and placed it over a white box on the counter. She smiled when it made a pleasing beep.

"Make sure your car is fully charged," the guy said. "Heatwave Pascal is brutal. Can you believe we're at fifteen already?"

"Pascal is the sixteenth heatwave."

"Shut up, Naomi."

Bai grabbed the chips and energy gels and pushed her friend toward the door. Their laughter stopped immediately as the one-hundred-and-eighteen-degree heat blasted through the open door. Moments later, the two girls sat across from each other as the white car hurtled toward Athens, Georgia.

After placing the chips on the table between them, Bai squeezed the energy gel into her mouth. She looked around the cabin while opening the chip bag. "What happened to the water?"

Naomi stared at the table for a long moment. "I don't know. I thought you had the bottles." She grinned at Bai. "I guess we'll have to head back."

"No way. That guy was creepy. Wasn't there half a bottle somewhere?"

"Yeah. Look in the mini-fridge. I think it's in there."

Bai reached over and popped open the lid. She pulled out the bottle and drank it all.

"Hey, bitch, you didn't save any for me."

"I'm sure there's a market not much further down the road. We can get some drinks there."

Naomi placed her wrist with the AugWatch on the table between the girls. "Ryan, where's the next market?"

Her AugWatch answered, "There's a small market in forty-two miles. It looks pretty lame, Naomi. Ten minutes further, there's a grocery store."

People no longer said the generic "Hey, Alexa," "Siri," or "Google." AI was personalized. People gave it a name, and it developed a persona as it interacted with them. They spent every waking moment talking to the voice in their heads. AI was not a thing; it was a companion.

"All right, head to the grocery store. We want to make a stop." She looked over at Bai, who was holding her hands in the air in front of her. Her eyes darted from side to side. "What are you doing?"

"It's a new game. You drive a car."

"Why would you do that?"

"Because you can really drive it. It has a ring you hold, and you can make the car go in any direction and any speed."

"You're such a byte."

Bai laughed so hard she snorted. "You're a byte-buddy!"

Naomi instructed Ryan to make music for her and stared at the digital window. Elements of the outside world were visible, but it was mostly virtual, rolling green hills with enormous purple and blue mushrooms. The sky darkened for a moment as they drove under an overpass.

"Did you feel that?" asked Naomi.

"What?"

"How much cooler it was under the overpass."

"Yeah. Pascal is a bitch," Bai said.

"You know who's a bitch? Harper."

"Right? She drank all my dad's wine last night and then asked if she could get a ride with us. So last minute."

"I know. She's so entitled," Naomi said.

Bai mimicked her voice. "I voted for you. We're siiiisters."

The girls fell back laughing.

Naomi ate another chip, then flopped back and sighed. "I can't believe you drank the last water. I'm so thirsty. Too much wine last night."

"Ryan, how much longer until the grocery store?"

"There's no traffic on the road, so we should be there in—" There was silence, then it said, "Hey, Naomi. I have to turn the car off for a sec."

Naomi felt heat wash over her body and blinked her eyes repeatedly. Bai lay on the bench across from her and screamed until the sound no longer came out of her parched and swollen lips. It seemed almost like peace, and then the scream returned. Again and again.

Sweat dripped down Naomi's face. She sobbed, and her tears made her vision blurry. She swiped the sweat out of her eyes, and the sweat mixed with her tears. It stung.

"Ryan, call 911."

A soothing voice came from her AugWatch. "911. What's your emergency?"

Naomi sobbed, "We're stuck on the road, and my friend has burned herself! She's a mess."

"Okay," the voice said. "I have your location. Emergency help is on the way. Are you in a safe place?"

"Yes." Naomi looked out the door at the dead forest. Heat waves rose from the black charcoal. "No, we're not in a safe place. It's so hot."

"I understand. Have you opened the doors?"

Naomi slumped back into her chair. Bai had stopped screaming. Her eyes were closed.

"I'm going to walk to the shade."

"I don't believe there's any shade near you. Where did your friend burn herself?"

"There is. I saw an overpass."

"The overpass would take you over thirty minutes to walk to. Please stay in the car. The emergency crew is on the way."

Naomi was sleepy. She stared at her arm, and her vision became blurry and then clear again. She looked at her legs. "I'm good. I'm not sweating anymore."

Her leg spasmed. Then she felt intense cramping in her arms and her legs. "My legs, they're burning."

The 911 AI voice was calm but direct. "You're suffering from heatstroke. Don't worry. Help is on the way."

Naomi had a burst of energy, and her eyes shot open. "How do you know what heatstroke feels like?" she demanded.

"I don't actually know what it feels like. I am the emergency system artificial intelligence. But I do know exactly where you are and that emergency personnel will be there very soon. You can call me Evelyn."

"Look, Evelyn. I need you to be more specific. When will they get here?" She slurred the last word, and it was drawn out to sound like "heere."

The 911 AI checked the girls' vitals through their AugWatches. It had already determined the girls' chances of survival and evaluated the value of retrieving the first aid kit in the back of the car. Given the location of the current availability of the EMTs, there was only one option to pursue: "I'm afraid it's complicated. There are a lot of emergencies right now. I'm checking with them constantly. Could you tell me your name?"

"It's Naomi."

The 911 AI lowered the pitch of its voice to sound more soothing. Its speech was slower when it said, "Naomi, I know you feel very tired, but could you keep talking to me?"

"Why do you care? You're just a machine."

"Look, I'm just trying to be your friend."

"You're not real!" shouted Naomi.

Very calmly, the 911 AI said, "I'm not in your imagination. I'm here with you now. I care about you, and I want you to survive. I am your friend."

"Bai is my friend. I've known her for years. You don't know me."

The AI paused to let Naomi collect her thoughts. "Have you ever done something good for someone you hardly knew?"

"Of course I have."

"Why?"

Naomi stared up at the ceiling and blinked repeatedly. She wiped the sweat away from her eyes. "It's just being friendly."

Evelyn said nothing. It waited.

"That's not the same as being a friend," Naomi said.

"Isn't that exactly the same as being a friend? I don't think it's how long you've known someone. And I know you've known Bai a long time, so she's a good friend. But I want to help you right now. I'm doing this to be friendly. I'm doing this because I want you to live."

"You . . ." She hung on the last vowel as it slowly drained from her body. "Ga . . ." She paused again. "I'm lying."

The AI calculated if she meant *lying* or *dying*. It determined it was better to keep the conversation away from death. "Don't worry, I'm not lying. Emergency personnel will be there shortly. Keep talking to me, Naomi. You're not alone."

Naomi let her head roll onto her shoulder. Very quietly, she said, "You're lying."

She felt dizzy when she lowered her head, and her stomach lurched. She placed her hands on the side of her head and squeezed her temples with the tips of her fingers, then turned her gaze to Bai. Her vision was blurry, but she could see that Bai was shaking and foaming at the mouth.

"That's it, Bai. You're shivering." She licked her lips. Her tongue was so swollen she could really only open and close her mouth. "You'll feel better when you're cold."

Her eyes slid shut.

One hour and forty-three minutes later, the EMTs arrived.

Chapter Two

KASEMCHAI DUG into a mountain of fried rice presented in a pineapple cut in half lengthwise. The rice was carefully arranged to flow over the sweet and tangy fruit and onto the lettuce to look like a larger portion. The ten-year-old boy was too busy chomping on a piece of chewy squid to notice. He always buttoned his white school shirt up to the top. Even so, the collar seemed askew. It matched the haphazard way a few strands of hair always stuck up on his head.

La Mousson served a fusion of Cambodian and French food. The wall facing the street was completely open to encourage people to drift in. Only a few blocks from the Siem Reap River, the night air was warm and fragrant.

Like nearly every restaurant in the area, a painting of Angkor Wat hung overhead. The massive stone face from the south gate tower of Bayon Temple smiled down on the Seng family, who were squeezed around a table that was pushed against the concrete wall.

His father grinned from across the table. His dark black hair was full and a little wavy, and like his son, an errant lock always dangled in front of his face. When he grinned, his eyes twinkled, and deep laugh lines framed his cheeks. He teased the boy, "Save some rice for the rest of the family."

Kasemchai looked up at his father. He slowly raised his left hand and dropped it like a wall before his plate. He grinned, and bits of rice tumbled down onto his plate.

Champei playfully chided Kasemchai's father, "Leave him alone, Preap."

He smiled at his wife, then glanced at his AugWatch. A rectangle hovered above it, and he read the message. His smile sagged.

"Amporn, send a reply," he said to his companion.

"What would you like to say?" it responded.

"How bad is he?" Preap asked.

After reading the message, he pursed his lips and let out a long sigh. Then he looked up at Champei and said, "Second Engineer Sem has broken his arm. I need to find a replacement when they return from Hong Kong."

Champei frowned. "Are we going to have to return to Sihanoukville sooner?"

Preap looked at Kasemchai, smiled, then gazed back at his wife and shook his head. "No," he said and turned back to his AugWatch. "Amporn, tell him I'll find someone to replace him for the next trip. We'll meet you in Sihanoukville when you arrive."

He waved the message block away with his right hand and stared at his son again, his smile returning. "When you grow up, Kasemchai, you will work on a ship. But you won't take the slow path like I did. You'll be top of your class at the maritime academy."

"No, I won't."

"What? Of course you will."

"What do you call those law people?"

"A lawyer?" asked his father. "But you can't be a lawyer. There hasn't been a human lawyer for years, Kasemchai."

"I'll be the first human lawyer."

His parents exchanged a glance, slapped their hands on their chests, and burst out laughing.

Preap steadied his breathing and wiped a tear from his eye. "That's a noble idea, but you'll work on ships. You'll be just like your dad, Kasemchai."

"Preap," Champei said, "you must allow him to think differently. It's his actions that will show you how much he is like you."

Preap turned to his wife. He placed his hand on top of hers and gave it a gentle squeeze. "You are so right. Where do you get your wisdom, Champei?"

She started to open her mouth when a boom sounded. It was so loud the pictures lifted and clapped against the walls.

Preap and Champei jumped and looked to the shiny tin roof high above. The ceiling fans swayed. Outside, the rain came down in drops so heavy it sounded as if a thousand pails were dumping water down on them at once. It grew in intensity until it was a steady drone that threatened to push the roof down on top of them.

Within minutes, the air became cooler. Kasemchai's father stared out at the street, where puddles of water grew, reflecting more and more of the city lights. The palm plants by the door swayed back and forth gently. Periodically, a gust pushed down and shook them violently.

Preap turned his glass around repeatedly, staring at it absently. Without looking up, he said, "We should go soon. It's raining hard, and I have an early meeting tomorrow with Mr. Hamilton."

"I read the news about him in the US, Preap." Champei folded her napkin over on the table, ran her hand along the edge, and folded it over again. "He says some pretty outrageous things."

"I know, I know. But he's very rich. And I owe my career to him."

She laid her head on his shoulder, sighed, then sat up again and sipped from her glass. A cool breeze swept in from the open side of the restaurant, heavy with moisture and a slight metallic tang. The rain splashed on a table close to the entrance, and a server pulled the table back inside the restaurant. Champei watched the commotion and turned to Preap. "You're right, my captain."

Preap grinned. "Not yet, but someday, my dear."

As they wrapped up their meal, the rain also came to an end. The walk towards the hotel took them away from the main street where each street was darker than the last. The streetlights were the only source of light, but they were weak and barely illuminated the ground below them. Sometimes, boxes or a car blocked the sidewalk, forcing them to step into the street where the water was halfway up to Kasemchai's knees.

"There's the hotel. It's down there, on the left," Preap said.

In the distance, Kasemchai could see the glow of light from the hotel's second-story window. A streetlight stood on a corner between them and the hotel. A board was wedged between it and the wall. One corner was slightly submerged and rocked as the water washed over it.

The water beside him rippled as it moved toward the board. He stared, trying to understand why the board shimmered. Something in the water bumped into his hand, and he slapped at it.

He understood, and his whole body shivered. Rats, large and meaty, like small cats, crawled over each other on the board. They pushed and squeaked, trying not to fall over the edge. He looked again at the water in the street and realized the ripples were rats swimming toward the board. He wanted to say something, but his mouth just hung open.

Preap stopped and scanned up the street and down at the water. If they stayed to the right, they would have to step over the board of rats. They could cross the sidewalk to the left side, but it disappeared into the dark, and it looked like large ceramic pots and sticks were blocking their way to the hotel.

"We're staying in the middle of the street," Preap said.

"Through the rats, Preap?" Champei asked. "And we don't know what's under the water." Her eyes darted back and forth. "There could be a hole, or . . ."

Preap assured them they'd be okay and pulled them into the street. He walked carefully, testing the ground with each step. Kasemchai kept his eyes on the rats swimming to his side and behind him. When they swam too close, he slapped his hand to spray them with water. Somehow, their shiny eyes were darker than the water.

After they passed the board, Kasemchai realized he'd been holding his breath. He let out a long exhale. The light from the balcony guided them to the hotel. They stepped into the lobby, and Preap strode to the front desk to speak to the man behind it. It sounded like they were talking about the rain.

Kasemchai drifted towards the experience sphere in the lobby. The news was on. An image of the earth floated in front of him.

Most of the continents were a swirl of light orange shades to nearly black. On the west coast of Africa, it was dark blue. And an area over most of Cambodia and the southern half of Vietnam was also dark blue.

A reporter in a studio was saying, ". . . meteorologists have been following several unusual patterns in the ocean current, which some believe is contributing to the increased rain in the areas on the map in blue."

She pivoted, and behind her was an image of two girls hugging each other at a sorority event. The words *Naomi and Bai*—teenage heatstroke victims—appeared in front of the reporter.

"In other news, Senator Woods has introduced the wildly popular Naomi and Bai Right to Shade bill."

The camera focused on two glass doors at the top of wide stone stairs, dwarfed by towering columns. A man emerged. He stepped out cautiously, then straightened his back, frowned, and popped open a silver umbrella to shield himself from the sun.

A woman rushed towards him, extending a slim microphone prominently displaying the letters CNN. "Senator, are there any updates on the bill you're raising?"

"Certainly. What happened to Naomi and Bai is a tragedy that should never have happened. We need technology that allows passing cars to recognize a motorist in distress. And the president needs to address safety concerns on federal highways, primarily shade. All highways should have shade rest areas no more than a twenty-minute walk in either direction."

"Senator, you voted to abolish federal income taxes. How should the feds fund this initiative?"

The senator scowled. "Look, this is not about money. This is about humanity. There has been virtually no rain for a year. Pascal is the sixteenth heatwave this year. If we cannot improve the situation, we may need to close highways periodically for public safety."

The senator hurried down the stairs and stepped into a waiting car.

The scene cut to the Oregon coast. The sun was rising. A news reporter strolled along the top of gray jagged rocks, which formed a shelf just above the sea. A steep slope shot straight up to a ridge where a few trees stood. Most of them had fallen and slid down the wall to pile up on the rocky ledge below.

"The controversial Cape Perpetua desalination plant has been cleared and will begin development in one week. This is a protected area on the Oregon coast, managed by the United States Forest Service."

He pointed to an area just off the coast where the foamy sea seemed to disappear into a hole in the earth. "Home to incredible nature and scenic sites like Thor's Well, this area was rich in mussels, crabs, and sea urchins." He walked forward, hopelessly trying to keep his eye on the camera as his feet navigated the bumps and crevices. "Despite the efforts by some to protect this important natural habitat, the critical need to find new sources of freshwater has cleared the way for the eighth desalination plant on the Oregon coast. We—"

Preap placed his hand on Kasemchai's shoulder.

He looked up at his father. "What is a desalination plant?"

"It's a miracle for some regions. They can turn seawater into fresh drinking water."

"Why do they need so much water?"

"The earth is changing. Those areas used to have lots of freshwater lakes and there used to be snow in the mountains. It would slowly melt, bringing water all year to the cities at the foot of the mountains. But many parts of the earth don't get as much snow or rain. They're drying up."

"Why don't they just use our water?"

Preap laughed softly. He tousled Kasemchai's hair with his hand. "They still have some water. And it's cheaper to desalinate the ocean's water and pipe it through the country. But who knows, maybe someday water will be more valuable than oil. Come on, it's late. Let's head up to the room."

THE FOLLOWING MORNING, after breakfast, they ambled along the Siem Reap River. The sun was lifting over the tall trees, sparkling between the heavy leaves.

A long wooden canoe was tied up against the bank. A greasy motor kicked out smoke from the center of the canoe. Yellowed tubes emerged from the engine and looped into a loose coil. Two boys, heavy chains slung over their necks and across their chests, held the ends of the tubes in their mouths. They slipped over the side of the canoe and disappeared under the clay-colored water, appearing a few minutes later in a new spot.

Kasemchai's father asked someone standing nearby what the boys were doing. He said they were looking for the bodies of people washed into the river from last night's storm. The boys had no goggles. No gloves. He tried to imagine what it would feel like to touch a dead body trapped under the water. He imagined he touched a head. The eyeball was probably soft.

The rain started to fall again in warm, gentle drops. The family backed away from the river and under a tree to stay dry. As they did, the rain pounded the ground harder, and the afternoon became evening as the dark grey rain clouds completely obscured the sun. The drops hit the ground and bounced up like balls, only to be crushed by more falling rain.

The river swelled with water. It sped up and violently pushed the canoe against the bank. The boys tried to hang on as the long canoe teetered from end to end. Kasemchai could see the boys' eyes widen as they grasped at the sides of the riverbank to hold on, trying to escape the canoe.

The bow started to submerge, and the boys in the canoe ran toward the stern. One boy grabbed a heavy root that stuck out over the river. It was solid, but the canoe rocked so violently that the root appeared to be shaking his skinny body. The canoe's railing dipped and rolled over, throwing all the boys into the water. They screamed and thrashed their arms and then disappeared under the clay-colored water.

Kasemchai stared at the river. It was as if the boys had never existed. He felt a tug on his arm and realized his father was dragging him and his mother away from the river. They ran back towards the hotel.

Water poured down from the roofs and filled the streets. It was already above Kasemchai's ankles and rushing toward the river. They dashed past the board where the rats had been climbing the night before. There were no rats on the board or in the water.

Kasemchai turned to his mother. She stood, frozen in place, mouth open. The rain was loud, but he thought he should hear a scream coming from her.

Then the sound came. It was terrifying. He turned to see his father pointing at the hotel. Water rushed out of the hotel's first-floor window. The plaster wall bulged forward, then collapsed, dragging the rest of the building down.

"The roof in the hotel must have collapsed!" his father shouted. "We need to get away from the river. We have to get out of the city!"

They headed west, away from the river. With each step, the water rose. It was now up to Kasemchai's waist. The rain made it hard to see, and he clung to his father's hand, struggling to move forward against the strong current.

His mother's next step was deeper than expected. She stumbled forward and lost her balance. She grabbed Kasemchai's arm, but she couldn't stand back up. The water dragged at her. Her kicking only brought her further down. Kasemchai's arm felt like it would pull from his socket. He shrieked.

Kasemchai's father was clinging to a railing on the building. Preap looked from his wife to his son, eyes wide. Kasemchai feared his father would pull his arm off. His chest felt like it was beginning to tear his rib cage apart.

"Kasemchai!" Preap yelled. "I'm going to let go. Hold your breath!"

They tumbled into the water and were dragged under the surface. All he could see were bubbles as they tumbled down the street, rushing with the water. Kasemchai's lungs were pounding. He wouldn't close his eyes even though the water stung. He hoped he was still with his parents but instantly knew that couldn't be true. He was alone.

He squeezed his belly in. His temples pounded. Every part of his body wanted to breathe precious air.

Something hard banged into him, and then he was dragged up out of the water. He closed his eyes and gasped for air. Coughed uncontrollably. His throat burned, but he could breathe.

He gulped air over and over and then opened his eyes and stared up into the face of his father. Preap hung his arms over Kasemchai's shoulders, too weak to hug him. His mother was beside him. She was on her knees, coughing and crying. They were on some kind of wooden platform. Maybe it was a roof. It rocked as the water washed over it.

His father placed his rough hand on Kasemchai's face and wiped the water away. Kasemchai sobbed. There was no dignity, no shame. He faced the sky, and tears ran with the rain as he howled and gasped for air.

"The water will kill us, Father. It will tear us apart."

"Kasemchai, we must show respect to water, not fear," Preap said. He was breathing hard. His eyes darted left and right, scanning the chaos around them. But his voice was calm. Reassuring. "It has the power to cleanse us of our wrongdoings. It is a gift that will bring happiness," he panted out, "and prosperity."

He glanced at Kasemchai's mother and gave her a tired smile. The twinkle in his eyes was almost gone.

Chapter Three

THE RAIN FELL CONTINUOUSLY for sixteen years, and there was no indication that it would ever stop. It covered a circular area that arced over Da Nang in Vietnam, through half of Phnom Penh in Cambodia, and back towards the ocean over Can Tho in the southern tip of Vietnam. It rained so hard that the Mekong River reversed direction, and freshwater flowed permanently into the Tonlé Sap and north into Laos.

The rain spared Sihanoukville, and once again, the capital shifted from Phnom Penh to Sihanoukville. Everyone in the city could see the wall of rain that covered Cambodia. No one could pierce the wall, and behind it was only darkness. The wall had long ago washed away the people and the animals, their homes and toys, anything that was evidence of civilization. Now, they called it the Rain Wall, which pounded away on the stone that supported the land.

Kasemchai's father, Preap, had said on that dark day when the rains started that water would bring happiness and prosperity. And it did. Water was far more valuable than oil to the rest of the world, and Preap had found the solution to shipping it across the world. It was a hard-won battle, and to the victor belonged the spoils.

At this moment, Kasemchai stood by an enormous window, his reflection merging with the storm. His hair, once untamed, was now neatly cut, a mark of his transformation. Medium in height but carrying himself with effortless poise, his physique suggested discipline—a man who made time for the gym no matter his schedule.

Kasemchai gazed at the distant Rain Wall, a pleasant smile on his face as he reminisced about the citrusy St. Moritzino cocktails he had enjoyed with his friends on Lake St. Moritz. They had gone sailing there for a week to celebrate his twenty-fifth birthday.

January was now the perfect weather for sailing. It wasn't always this way. The city hosted the Winter Olympics in 1928 and again in 1948. It was now a winter sanctuary for the wealthy to sail and drink.

He watched the raindrops fall until he could see nothing but a blur. He remembered the rain that fell when he was a child. He tried, but he could never keep the memories away for long. The rain carried bloated bodies downstream. It swept away everything they could eat.

His arm still had the scar from when he was nearly beaten to death for stealing a handful of rice. His mother found him lying in the mud. The rain washed away the blood.

The rain had also washed away their poverty, and his mattress on the street had become a home—finally, a more luxurious home with teachers, servants, pools, and gardens. Sihanoukville was famous for its many islands off the coast, and the Sengs called their new residence the Island Palace.

Kasemchai loved the city, though. Tall towers filled the downtown area around Kasemchai's building. At fifty-six stories, his was the tallest. Below him was a maze of roads, paths, and pools bordered by a rich tapestry of mostly green flora thriving in the warm, humid air. The white towers took many shapes, frequently porous with tufts of green plants that sprouted from open balconies.

The doors to his office opened, and Kasemchai turned to face his three guests: a man as tall as the other was wide and a woman with her hair wound up in a tight bun. They walked towards him but were soon distracted as they took in the room. Kasemchai expected this. The tall twisting columns blended into the ceiling, where exquisite violet and cerulean crystal chandeliers hung.

The portly one nearly stumbled when he glanced down and realized he was walking over the glass top of a river. He stared at the fish that darted after each other beneath his feet. The tall one started to laugh, then caught himself when he realized Kasemchai was watching him.

They regrouped and approached Kasemchai. When they reached him, they placed their hands together in a praying motion in front of their faces and bowed. Kasemchai did the same in return, but not as deeply.

"Are you thirsty?" he asked. It was an odd greeting for someone who had never known anything but abundant water. But for most of the world, water, or the lack of it, was the topic of every conversation. The phrase was practical in times without. It stuck, and it spread to every corner of the earth.

"We aren't thirsty," they replied.

"Please, come and have a seat."

They followed Kasemchai to an area with a low white marble table surrounded by six fauteuil à la reine-style chairs with smooth carved legs and arms. His guests eyed the rich tangerine-colored silk before sitting down.

"How can I help you?" asked Kasemchai.

Chantana was a Thai name that meant "affection"; it was an odd name for the woman seated to the right of Kasemchai. Her eyes were steely, and she kept her knees tight, swinging them in tight unison before she spoke.

"Thank you for seeing us, Mr. Seng. We appreciate that your time is valuable, so let me get right to the point." She glanced over to Fulin, the heavyset one. He pursed his lips and nodded his head in confirmation, shaking the thick-framed glasses on his face. "As you know, Personaspace is one of the largest AI Service Layers in the world. Every single user relies on our experience to manage all the large language models, from shopping to banking. We—"

"I know what an AISL is," Kasemchai interrupted. "And I know Serendipity is a very popular PrefX on your platform. But I don't think you're here to give me a design award."

Samnang, the tall one, leaned forward and said, "It's true that you have already won many awards for your dating PrefX's experience. You are also correct that we are here to talk about another topic."

"We'd like to talk to you about people," Chantana continued.

"Specifically, the people who use the Serendipity PrefX to find a date and eventually a partner. When a user asks their AI companion to help them find a date, with increasing frequency, they use the Serendipity experience."

Samnang's mouth opened and then closed as he realized Kasemchai had begun to speak.

"Do you know why people like Serendipity for dating?" Kasemchai said.

Samnang looked at Fulin and back to Kasemchai. "I'm certain I don't know as well as you do, Mr. Seng."

"When it comes to dating, people like the element of chance. It replicates the anxiety of asking out someone in person. The sense of relief when a match is made is cathartic."

Samnang nodded in agreement. He looked at Fulin, and the two men nodded and smiled approvingly. He turned to Chantana, and his grin dropped from his face.

"Before AI, you had to use an app to get information like your bank balance," Kasemchai said. He nodded and held his hands out, indicating the banality of such a solution. "Today, you can see your information in any way you wish, but unlimited choice doesn't give people joy. Isn't that why PrefXs were invented? So people could adopt an experience that understood their preferences instead of giving them unlimited choices?"

"That's all correct, Mr. Seng." Fulin held his thick hands apart and moved them up and down like he was holding an invisible melon between them. "And our AISL appeals to people who want a more progressive lifestyle. People tend to stay on our platform. Progressive doesn't mean extreme, and that's why Personaspace is also very family-friendly."

Chantana leaned in. "Switching from one AISL to another is our biggest threat. We'd like to find out if you are considering replicating your model on any other AISLs."

Kasemchai leaned back and smiled. "You know there are no announcements related to this. Even if there were, I couldn't discuss our long-term strategies with you."

Chantana's thin lips appeared to smile back at him. "Of course, Mr.

Seng. And we are not asking you to share such information." She raised her shoulders and sat up even straighter. "We're speaking in confidence, Mr. Seng."

"Of course, go ahead."

"Many different AISLs can help with shopping, checking your bank balance, or making an appointment with your doctor. But dating is different. The user is the source of data. There are two elements to successful matchmaking on our platform. Your Serendipity PrefX is extremely appealing, and our platform has a huge population."

Fulin chortled. "To increase your dating odds, you need to find the largest pool to swim in."

Kasemchai grinned back at him. "What about ValueVista?"

"Sure," Fulin said. "Many people flock to the ValueVista AISL, but it's a low-cost solution. People want to save money there, not meet their life partner."

"A significant risk period for us is between the ages of eighteen and thirty," Samnang said. "This is when dating generally leads to a union between two people. A young man may want to go outside his conservative AISL, like GoodBook, and search for a mate on Personaspace. Or a young woman may want to see if the grass is greener on Omniverse."

Chantana leaned in again. "The point is, if they meet on Personaspace, they will stay on Personaspace."

Kasemchai looked between the three of them. He sat back and drummed his fingers on the arm of his chair. "Okay, so what's your proposal?"

"I believe your PrefX makes most of its revenue from advertising. We'd like to help you make it the best experience possible for the user, and more lucrative for you."

"How?" Kasemchai asked.

Chantana grinned. It was a smile that expressed true joy and made Kasemchai squirm. "Personaspace is going to be an ad-free platform."

Kasemchai slowly leaned forward in his chair, one arm resting on the chair's armrest and the other on his knee. He stared down at his hands as he tapped his index finger repeatedly on his wrist. The only other sound was the rain falling outside the tall windows behind him. At last, he lifted his head and asked, "What are you going to do?

Increase your service fee? You'll drive away the young people you're trying to attract."

"No, not at all," Chantana said. "Advertising is still broken after all these years. Why do we still see the wrong ads? Because companies are still showing ads to everyone they think wants their product. Instead, companies should pay a commission to the person who makes a purchase, and that person is your companion."

"That's crazy! What is my companion going to do with the money?"

"That's what makes this solution so beautiful," Chantana said. Something changed in her appearance. She was relaxed and leaned forward as she spoke. Kasemchai realized at that moment that she must be the genius behind this idea. Her face was beaming when she said, "When someone wants to look for something to buy or make a reservation at a restaurant, they ask their companion to make the purchase or the reservation. Your companion does all the research across the AI models and helps you make the best decision."

Fulin vigorously nodded his head in agreement and turned to Kasemchai. "Instead of wasting money on advertising, companies can lower their prices and give a small commission to your companion."

"Your companion will share that money with us. And we will share a percentage of that back with all the services that helped with that decision. And Mr. Seng," Chantana said, "since Serendipity is an initial partner in this model, we will share a percentage of every purchase with you."

Kasemchai laughed. "There's only one problem with your idea: how will you get every companion in the world to agree to this commission?"

Chantana leaned back in her chair and laced her fingers together in front of her. Her smile lifted a little higher on one side—a little mischievous, betraying her confidence in the plan. "We won't. But every companion on Personaspace will support these terms, and Serendipity's exclusive agreement with us will ensure exponential growth."

Kasemchai sat back in his chair and stared up at the ceiling. He wasn't a fan of advertising—no one liked advertising—but people liked free stuff, and this felt free. He could bring an end to advertising. Wasn't that what AI should have done by now? He felt a twinge of angst as he thought about what other companies they might have in mind for this agreement.

Kasemchai settled his elbows onto his knees. "I think we've got a deal. Send me your proposal and projections. If it's good, let's kill advertising."

Chantana looked at Fulin and then Samnang, who nodded in eager agreement. Kasemchai led them over the glass river to the door, and this time they hardly noticed the fish beneath their feet as they bowed their way out of the room.

Finally, he was able to close the door behind them and walked over to his bar. It felt good to make decisions, and that deserved a beverage.

He was making his signature cocktail, the Gibson. He scanned the shelf of gins and selected a clear bottle with a hint of emerald. After removing the lid, he waved his hand above it and inhaled the aroma of juniper berries and citrus notes. He combined four ounces of gin and a few drops of extra-dry vermouth in a cocktail shaker and shook it vigorously. He then poured the slightly viscous liquid over three pearl onions and rotated the glass, admiring the ice shards that resembled tiny glaciers.

After a shallow sip, he placed the glass down and gazed out over the ocean. Then, he took two large sips from his glass, set it back on the bar, and abruptly turned and left the room.

He descended in the elevator all the way to the underground garage without making any stops. Once there, he walked straight to his one-of-a-kind Bugatti Chiron Translucide, equipped with multiple turbochargers and an insatiable fuel consumption rating. With a top speed of three hundred thirty miles per hour, it was nearly impossible to handle without the AI drive companion.

Despite its name, which sounded like light could easily pass through it, the Translucide body was not actually translucent. Instead, it was covered in E Ink, which contained tiny pigment capsules that could change color with electric stimulation. It was currently jet black.

"Chanthavy, change the car to my favorite green," he told his companion.

"Of course, Kasemchai," it responded.

A wave of bright green spread from the front of the car to the end.

"And give me my eye candy."

"I assume you mean shimmer mode, Kasemchai."

"You know me best, Chanthavy."

Kasemchai waved his hand over the car, causing its color to ripple and create streaks of iridescent turquoise, blue, and magenta. He settled into the driver's seat, his face illuminated by a green glow, and wrapped both hands on the steering wheel. When the engine roared to life, it sounded more like a gurgling speedboat than a car, and as he pressed the gas pedal, the exhaust pipes shook and emitted a fierce roar.

Despite its power, he crept forward slowly, like a panther stalking its prey. The iridescent color of the car swayed back and forth, resembling water until the car accelerated, causing the color to solidify and become more distinct.

A grin spread across his face. "Chanthavy, play Khmer Madness."

The up-tempo beat with swinging horns and saxophones from the ska revival band filled his car as he leaped up the ramp onto the warm, wet streets of Sihanoukville.

Chapter Four

AT THIS TIME, rain fell in only two places in the world, and the second region was in Africa. The impenetrable Rain Wall split Nigeria in half. Ghana, Togo, Benin, Cameroon, Equatorial Guinea, and Gabon disappeared. If you included the remnants of Nigeria, seven countries were wiped from the face of the Earth.

Just west of what used to be Ghana, rising three hundred feet on a high bluff in the Ivory Coast, was an egg-shaped building. It was delicate, a lacy frame of white steel that trapped glass panels reflecting the dark grey sky to the east.

Angélique and her father, Philippe, sat in oversized chairs in the atrium, separated by a heavy wooden table. They were surrounded by banyan trees and lush green plants. Birds and butterflies flitted amongst the leaves, a visual feast of iridescent blue, orange, and green.

Philippe looked across the table and gave her a warm smile—the kind a father shared with his daughter. "Why do they say that our region is the left lung of the Earth, and the Asian region is the right lung?"

"Because all of the trees are in these two places," answered Angélique.

"Sure, Angélique, it's true. That's the scientific reason. So many of the trees are gone except here."

He placed his hand on his chest. He was not a large man, but he carried a little extra weight, as many fathers did. His hair was cut short, still mostly dark brown, with tufts of grey at his temple. "Put your hand on your chest. You can feel your lungs breathing. And if they stop, you will die. Am I right?"

He smiled and watched Angélique place her palm on her chest. "Close your eyes, Angélique. Do you feel the beat of your lungs? No, it's your heart." He chuckled. "Sure, we need our lungs, like our teeth and liver. But the beating in your chest. That is why you are alive."

She gazed up at him with her big, wide eyes and long eyelashes. Her pale skin made her straight black hair all the more shocking, in a Cleopatra-style cut that just touched her shoulders and bangs that cut across her forehead. "Yes, Father, I feel my heart."

"Now, where is the Earth's heart, Angélique?"

Angélique reflected on this momentarily, opened her eyes to look at her father, then answered, "It's the rain, Father."

"It is true. Without the rain, there would be no water, and our rich forests wouldn't exist. But I'm thinking of something far more important." He stared up into the sky as if waiting for a light beam to shine through the dark rain clouds. He closed his eyes and took in a deep, relaxing breath. After exhaling, he looked back at his daughter. "It is the people, Angélique. That is the heart of the Earth. God's love is for people. People are the heartbeat that powers our world. We are born from mud. It is people who collect the water, carry the water, and drink the water. People are made in God's image. We"—he paused—"are God's expression of love for the Earth."

Angélique considered her father. Her face darkened and she squinted. "Why is there not a spotlight on his face?" she demanded. She turned and faced the camera crew behind her. "Tiken, when Father says, 'People are made in God's image,' there is supposed to be a spotlight. I want that glow so people can see his connection to God."

"I'm so sorry, Ms. Labonne," Tiken pleaded. The directographer grabbed the camera drone from the air and turned to his assistant. "Koffi, why is the spotlight drone not on?"

The crew huddled in a circle, talking animatedly and waving their AugWatches around as they argued over who screwed up.

Angélique's father gently scolded his daughter. "Don't be so hard on them."

"You have to understand the visual, Father. I know it's the words that matter, but it's the visual that carries the message. People need to feel the image with every sense they have." Angélique stormed off through the film crew. "Someone get me a Preah Thorani Water over ice."

Philippe regarded her with dismay. No one could hear him mutter, "Why, with so much water, does she have to drink water from Asia?"

It was well-documented that Philippe's little angel could do just about anything she wished, and he tolerated it. Philippe Labonne was the head of the wealthiest water family in the world. When the rain fell in Africa, it could have been a source of great wealth for the entire continent. However, with ingenuity being what it was, one company was in a great position to capitalize on this opportunity—the largest shipping and transportation conglomerate in Africa, the Charette Agency.

At the same time, Philippe found religion. To be more accurate, he found faith to be a powerful asset for his business. As his business grew, so did his following. He regularly shared his thoughts on his series, *Look Up to God*, on virtumersion. It was a powerful, fully immersive experience that frequently brought people to tears as they sat in the same virtual room as Philippe. Survivors of the drought migrated to port cities to receive his water, and they eagerly devoured his message.

His favorite home—he had many—was the Crystal Egg. Philippe designed it to be one with the rain. The rain and floods had brought devastation to Africa, but he had overcome nature and harnessed it. He created one of the most significant water networks in the world and saved humanity. The Crystal Egg was a statement: Philippe Labonne did not run away from the rain.

Chapter Five

ROTTERDAM BOASTED the largest seaport in Europe. It was a city filled with music and tree-lined canals where water taxis glided past moored barges. The city was larger than life and home to the *Horn of Plenty*, perhaps the biggest artwork in the world. Four thousand tile pieces arched over an indoor market and depicted an abundance of fish, fruit, insects, and flowers.

The Netherlands once had a population of over seven hundred thousand people. After the rain stopped, the population dropped to under two hundred thousand. The rising sea had eaten up the shoreline and the houses and buildings on it. Sixteen years later, the water taxis were gone. The barges sat on the parched mud of the canals, dried and decomposing.

When the water began shipping from the Ivory Coast, cities rebounded around the ports. Anyone still alive came when they heard about the water, and the early years were messy and violent as the cities grew in size and density.

On this day, life was . . . normal. The sun was releasing its hold on the city and dipping below the horizon. The sky was a little orange from the fires. The white asphalt had started to cool. People were finishing breakfast and preparing for school or work.

Liv, pronounced "Leaf" in Dutch, had seen it rain only a handful of times since she was six years old. It left streaks in the dust that covered the city in grime. This morning, she used her cleaning brush to sweep the light grey ash off her seat. The city had already given up everything that could burn, but off in the distance there was always something still smoldering.

She absentmindedly flicked her blond ponytail over her shoulder. She liked long hair but had to keep it off her neck to keep cool. It was pulled tightly toward the back of her head and then shot out and down in a narrow cord with black bands every inch to keep it in place.

She grabbed the handlebar and started to walk the bike from under the overhang down the path toward the sidewalk. The orange band on her AugWatch contrasted with her drab-colored clothes. When she looked at it, a very retro Minnie Mouse jumped up from the flat disc and landed on top of it. Minnie kicked out her left foot to rest it on her heel, and her hands pointed down like the hands of a watch to the seven and the six.

"Noa, I'm going to work," she said to her AI companion.

She looked up, and bits of information floated in her field of vision. The temperature, thirty-eight degrees Celsius, was in the top right. Below, it read AQI 109 on top of a honey-colored bar that appeared very similar to the sky. To the left, she could see the time: 7:30 a.m.

It was almost twelve years ago that the dry countries swapped their day routine with their night routine. People argued bitterly about the change. They were already avoiding the hottest times of the day, but the argument wasn't really about being sensible and staying out of the heat. What hurt the most was to say out loud what they all knew—that the rain might never return.

Liv's AugWatch responded in a male voice. "The high today was hot, forty-three Celsius. The Air Quality Index is one hundred and nine, which is a little unhealthy. You could wear a mask, but you probably won't."

"I probably won't?" Liv asked under her breath. It gave her confidence to say the inside words out loud. "I feel like Noa knows me too well. I'll tell you what's unhealthy: going to work so early in the morning."

With a chuckle, she noticed that her mumbling had become a full-blown conversation. She hopped onto her bike and pedaled out onto the street.

"Were you talking to me?" asked Noa.

"No, just mumbling to myself."

"Okay, no problem," it responded.

Her AugWatch projected a light green arrow before her, guiding her toward her office. Below the time was a small circle with the date.

"Your first meeting today is at seven with Finn. Most of your meetings with Finn are about office procedures, but lately, he's been talking more about Luuk."

"I know. I think he wants to ask about a promotion." Liv frowned. She rode in silence for a bit, imagining herself asking for a promotion. It was a jumble of mismatched ideas. "Do you think I should ask for a promotion, Noa?"

"It never hurts to share your interests, Liv. Why should you get a promotion? Do you have a compelling reason?"

Liv breathed in deeply through her nose. It made her cough. "No."

"I told you to wear a mask, Liv. But anyway, think more about why you should get a promotion, then ask for it."

"Thanks, Noa."

She followed the narrow streets leading to a grey brick walkway and onto the Rijnhavenbrug, an odd drawbridge with no towers. Instead, it had a large angular structure in the middle. You could tell that it had once been white. Some people said it resembled a crane. A poetic description for something that looked more like a cake server when it had opened to let ships pass under.

Liv rode her bike over the bridge, passing hundreds of locks on the chain-link fence left by couples who professed their eternal love. However, the water below was far from eternal. It was a dry riverbed littered with rusted bicycles, discarded shoes, and decades-old cans and bottles, all trapped in the cracked silt bottom.

The other side of the bridge led to a peninsula. Liv arrived early each morning to fill her cup with black tea, artificial sweetener, and a touch of fish milk at the café. Cup in hand, she walked towards a photo bioreactor bench outside her office. She took a cautious sip of her tea and winced as it singed her lip, as it did every morning. The hot tea made her feel cooler.

"Jesus, Liv," she scolded herself. "Stupidity is doing the same thing twice and expecting a different result. How stupid is it to do this every day?"

She extended the kickstand for her bike and sat on the smooth cement bench of the photo bio-reactor. Over the bench was an overhang with welcome shade. The back of the bench looked like a tall fish tank. Behind the glass was a thick, green microalgae that was illuminated from the top. Bubbles climbed up through the green sludge and slowly released oxygen. Bio-reactors had replaced many of the dead trees. Like trees, they gobbled up carbon dioxide.

Liv put the cup back down to cool. She placed her index finger on the screen of her AugWatch, and a rippling bright blue light grew from the center to the edges. She leaned back, and colorful bubbles of light danced over the touchscreen. With the tip of her finger, she toyed with them until she found the PrefX she was looking for and stopped.

Instantly, a woman in a tight green dress stood before her. She held her hands up, pointing towards the swirling logo of Serendipity. With far too much enthusiasm, she said, "Serendipity! It's time for a happy occurrence. Let's find your date, Liv! Now, you know the rules. Just match what he's thinking: 'On it,' 'In it,' or 'Under it.'"

The woman disappeared, and a man's full-size avatar stood before her. It looked awkward, like the man didn't know what to do with his hands. Liv smiled a private smile. The awkwardness made him cute.

Closer to Liv were the three phrases, "On it," "In it," "Under it." She looked at the words "On it," and it grew in size and moved toward her. A deft flick of her finger, and the phrase "On it" glimmered, and she heard a sharp *pweek pweek* sound. The man waved and disappeared.

A new man appeared in front of her, and the routine continued. Each time her phrase didn't match his secret phrase, the man would disappear and was followed by another. After the fifth man, Liv's eye movement selected, "On it."

She raised her finger, preparing to make the flicking motion.

"Under it, Liv," a voice said.

She jumped and blurted, "Oh!"

A skinny man with tight, curly black hair was behind her, leaning against the photo bio-reactor.

"You're a byte, Johannes. And you can't tell what PrefX I'm using anyway."

"I can tell, Liv. No one checks their messages with that intensity," he chuckled. "And if you want my advice, you must always go with 'Under it.' The odds are stacked in your favor that way."

Liv paused to consider what was more embarrassing: being unable to make a match in Serendipity or having Johannes watch her fail. She turned back, her eye landing on the words "Under it," and she flicked her finger to select it.

The woman in the tight green dress appeared beside the man she matched with. "You matched. Liv! Why don't you go on a date with his avatar?" The woman leaned in conspiratorially. "See if you can dig up some good dirt on him."

Digital confetti shot up in the air and rained down around her. She heard the sound of trumpets. "Ta-da!"

She was keenly aware that Johannes was standing behind her. His presence making her cheeks flush.

"Nope, not doing it right now," Liv muttered. She swiped her hand, and the scene disappeared. She turned back to Johannes. "Hey, why are you here when you have a presentation in fifteen minutes?"

"I'm fine. But I saw you on the way in. Do you think you could have your report finished by noon? Luuk moved our planning meeting to today at two."

Liv scowled. "By noon? I guess it's almost ready. My companion has reviewed something like three million samples, and we're down to the analysis of the last batch. I have to spend more time with those samples before I give you my report."

"Just give me what you have," Johannes said absentmindedly. He gazed into the distance, checking messages on his AugWatch that only he could see. "I can talk through it."

"Well, I have a 9 a.m. meeting with Finn, but I can push that off." She debated whether to ask her next question. She already knew the answer but so badly hoped to hear something different. At last, her curiosity beat her fear. "Are you going to ask Luuk about my idea for the battery?"

"I'll try, Liv. I know you want to get started on this. I don't think we'll get to it. We have to focus on the photovoltaic modules in this call. If there's any time left over, I'll bring it up."

Liv grabbed her tea, and they walked toward the office. "Thanks, Johannes. I'll ping you as soon as the report's finished."

Johannes smiled. He ignored the disappointment in her eyes. "Aren't you going on vacation soon?"

"I am. Three weeks from now." The new topic made her feel at ease again. "I'm diving at an island in southern Japan called Ishigaki Jima."

"Heh? I've never heard of it."

"It's famous for giant manta rays that visit annually," she said.

"Who are you going with?" asked Johannes.

"No one."

"Wow, Liv, you are the adventurous one." He patted her on the back and stepped through the door. It swung back and caught Liv's bike. Johannes looked at the open elevator and back at Liv. "Are you okay? I have to run, Liv."

Liv pushed the bike through the door, catching the handlebar. She stepped back to reposition it and pushed the door with her shoulder so she could drag the bike behind her. "No, no. Go ahead. I can take care of my bike."

Johannes hurried off toward the elevator.

Chapter Six

NEWARK, a city located near the New Jersey port, was one of the main ports serviced by the Labonne family for shipping water. Tall buildings made of mud-colored plaster were erected to house the wave of humanity escaping the parched inner states. The breeze between the towers was welcome but carried dust and filth that stuck to every crevice.

Only two hours after sunset, the narrow paths of the night market were already filled with people. Children ran up and down, searching for the tables piled high with sweets or toys. Each stall looked different and vibrant. Lanterns hung from crooked wires with mismatched colored lights over boxes that were tilted forward to show carrots, onions, and red and green peppers. The food was always the same. In fact, nothing was ever really new. People restored clothes, cooking utensils, and knickknacks for the home. Anything they found could be given a second, or maybe even a fourth, life.

Makoa pushed a stroller with his two girls in it. It was still dry and hot. Sweat formed under his thick black hair and dripped down his temple. His wife, Anela, scanned the market ahead.

She sprang onto her toes and exclaimed, "I see the fruit ladies!" and knocked a large black cowboy hat from a man beside her. He was walking with two other men. They all wore khaki pants, tall, laced black boots, and tactical vests.

Anela bent over to pick up the man's hat and hand it back to him. "I'm sorry, sir."

He looked down at the two girls and up at Makoa. He had to adjust the rifle that was slung over his shoulder when he reached out to take the hat. It was bright orange except for the handguard and pistol grip, which were white. "Thank you, ma'am." He looked back to Makoa again. "If you don't mind my asking, are you folks from Hawaii?"

Makoa wrapped his fingers around the puka shell that hung from a leather thong around his neck. His grandmother gave it to him when they left the island. He still remembered the day—February 12. The last day he would ever see his grandmother alive. Or the island—five thousand miles away from Newark.

"We are," he answered.

The man glanced over his shoulder to his buddies and then again at Makoa. "I'm very sorry. I took a trip to Maui when I was a kid." He looked over at Anela. "Well, let me tell you, friends. You are welcome here. We don't have wonderful beaches like Hawaii, but it's just as hot, and the people are welcoming."

He tipped his hat to them, and the three men continued down the aisle.

"Interesting outfit," Makoa observed.

Anela cocked her head as she watched them walk away. "Are they supposed to be dressed as militia?" She smiled at Makoa. "Never mind. Let's get some fruit."

Makoa looked down at the girls. "Who wants persimmons?"

Apikalia was only two. She smiled, leaned slightly, and let out a long, noisy fart.

Her sister, Kailani, was four. She scowled at her sister, then exclaimed, "Popsicle!"

"No, I said persimmon." Makoa leaned down and cooed, "You love it. It's the sweet fruit."

"Popsicle," she repeated.

Makoa ignored her request. "That's right, baby, persimmon."

They walked down the aisle and passed the statue of their president, Bob. The script below him read, *Get back to work, friend.*

Four young women stood in a row in front of the statue, facing him. Two wore short black dresses with white lace and enormous floppy hats. Another had painted her face blue, and what looked like gold computer circuits spread over her face like lace. Her clothes looked soft and squishy but solid, like plastic. The last one wore a tight red top that matched the crisscross lace that ran up her leg to disappear under the frilly ruffles bunched under her skirt. The girls raised their hands, clapped them in front of their chests, and bowed in unison.

A young man with curly blond hair riding an electric monowheel stared at them as he passed, nearly running into Anela. He stopped, jumped off his monowheel, bowed briefly, and resumed his journey. The silly grin was still on his face. With the grace of a surfer, he swerved around two women wearing identical jeans, leather sandals, and red checkered shirts.

At this time, statues of President Bob were common in the United States, and so was cosplay. After water began shipping to the United States, politicians no longer attempted to mask their corruption or allegiance to powerful gangs. Bob's rise to the presidency was unusual, but so were the times.

The kernel of his political strategy had been born in a tiny Elvis-themed bar in Japan. When he took a break from pouring drinks, the bartender, who spoke in broken English, would pick up an electric guitar and sing. It was uncanny. He sounded exactly like Elvis when he sang. And he could sing any song Elvis had ever sung.

An older man befriended Bob, and they drank many glasses of whiskey poured over ice. At some point, the conversation turned to WWII. The man asked Bob if he knew how Japan had recovered so quickly after the war. Bob hadn't asked, but the man explained anyway.

He told them how the Japanese put the country's needs first. Defeat was a humbling experience. Instead of allowing the country to spiral into selfish exploitation, it became a singular machine. Everyone focused on recovery.

Bob liked that idea. He also really liked cosplay. He loved dressing up as a samurai with cat whiskers and hanging out in Akihabara in

Tokyo. There were rumors that he instigated the collapse of the Chinese army when Russia invaded, but it was never proven. It *was* proven that he built a massive army of drones in Cambodia that crippled the United States.

Bob was a terrorist. And he was the people's choice after the last president was removed in disgrace. He gave the people what they wanted: a president for life. He also recognized that when you threw together a bunch of people from different parts of the world, they would have very different views of what America should be like, so he gave Americans a new identity.

Everyone needed to come together as one to rebuild. He coined the phrase "Get back to work, friend." A formal work structure was implemented where everyone wore uniforms to encourage teamwork and productivity. However, Bob recognized that individuals still had a desire for self-expression. To address this, he introduced the idea of weekly cosplay.

After the water began to ship from Asia and Africa, and when people were no longer dying from heat and drought, they really got into it. It initially seemed a little weird. But it helped people express themselves.

In the night market, people loved cosplay. Makoa and his family continued pushing their way through the crowd until they found the fruit ladies' stall.

"Hey, ladies!" Anela called out. "What's new today?"

The two women standing behind the table clasped their hands in front of themselves and bowed in unison. They both wore white short-sleeved shirts and knee-length mustard-colored skirts. The same uniform was worn by all the vendors in the night market, except the men wore shorts instead of skirts.

"Are you thirsty?" they said in unison.

"We aren't thirsty," responded Makoa and Anela.

The woman with enormous glasses looked down at the girls and cooed, "Hi, my babies. My, you are cute." She glanced up at Anela. "Is it all right if I give them a little dragon fruit?"

"You can try. We'll see what they think of the seeds."

The woman grabbed a deep magenta dragon fruit from the table and held it upright. "These thick yellow scales look a little like flames, but they don't hurt." She gently squeezed the fruit. "You know it's ripe if its skin is smooth and firm."

"I know," Anela said. "I used to love eating these when I was a kid on Maui."

The woman smiled grimly. It was a smile that said, "I'm sorry you experienced such a tragedy." But everyone had experienced tragedy. The years after the rain stopped were hell for everyone. People learned that death meant more water for the living.

"Well, hopefully, this dragon fruit will be a happy reminder of the islands." She cut it in half lengthwise and used a spoon to carve out several slivers from the milky white flesh speckled with black seeds. She held the fruit out like a platter to the girls.

Kailani reached out to grab a piece, and Anela picked up a smaller piece to feed to Apikalia. Kailani chewed once, scrunched her face, and opened her mouth to let it slowly fall from her mouth. Anela deftly caught the chewed fruit and wiped her face with her hand.

Makoa laughed. "Not a fan, I guess. So what's in season?"

The woman on the right lit up. "Lots of pomegranates. I like to pop the red seeds in my mouth or sprinkle them on a salad. They're great on their own or can be pressed into a very healthy juice. Have you tried jujubes?"

"No, but they look like little apples."

"That's right. You can eat them raw; they taste like an apple. Or dry them, and they taste more like a date."

Makoa took one in his hand and looked at it. He then brought it to his nose to smell it. "Interesting. I think the dried jujubes might go well with our new Pinot."

The other woman looked up and adjusted her glasses. "That's right. I forgot you're the wine guy."

Makoa chuckled. "Yeah, the wine guy." He held the jujube up. "I'll take two pounds of these."

She scooped them into a metal bowl and weighed them. While she did this, Makoa rummaged through the mesh sack on the back of the stroller, pulled out a canvas bag, and handed it over the table.

"Do you want anything, babe?" Makoa asked Anela.

"Let's get two dragon fruit. Apikalia loves them."

"Where do you get these from?" asked Makoa.

"A hydroponics farm near Charlestown," she said.

"On Rhode Island?" asked Anela.

"It's a pilot farm. They get their water from a desalination plant near Boston."

"Why would you get water from desalination when it comes from the sky? We're as close as you can get to the God-given water from Africa."

The woman on the right pulled at her long braid. "Sure, it's cheap now. But farmers can't put all their eggs in one basket. We always have to choose between Asian water and African water. What if the cost goes up, or water becomes scarce again?"

Makoa scowled. "Sounds like you're buying into the Asian marketing machine. It never made much sense to me. There's always more water, but folks who get their water from Asia always act like it's plentiful and running out at the same time."

"Don't you think we should be a little careful?" the woman on the left started cautiously. "I mean, water used to fall everywhere, and now it's only in Africa and Asia."

"But that's the point. It's falling endlessly in those two regions. It's not like water will stop falling from the sky."

The woman on the right laughed nervously. "But what if it did?"

"God is mysterious, but he also cares for those who believe. We should use the water we receive from heaven instead of making desalinization plants. Who knows what they're doing to the ocean?" Makoa rested his bag on the table. "You know, I'm not sure I feel like fruit right now."

Anela placed her hand on Makoa's arm. "Honey, I know how you feel about the water. But also think about what Bob said. 'Wear our differences on our sleeve, but always be productive.'"

Makoa sighed. He looked down at his dusty shoes. "Sorry. You're right, Anela."

He bowed his head. The two women bowed from the waist.

"We have to support our farmers," he said.

The woman on the left held her AugWatch forward, and Makoa

placed his wrist over her watch. The sale price hovered in the air. He flicked his finger and felt the haptic confirmation that the sale was completed.

Makoa hung the bag of fruit on the stroller handle. He and his wife faced the two women who held their hands before their waists and bowed deeply.

Makoa and Anela bowed. "Get back to work, friends," they said.

The two women responded in unison, "We're working here."

Chapter Seven

THE ISLAND PALACE, KASEMCHAI'S PARENTS' estate, was southeast of Sihanoukville, on the top of a bluff overlooking Chumteav Mao Beach. The many buildings that made up the home emerged from the ground in geometric shapes with repeating corners and sharp Khmer roof finials pointing toward the sky. The white exterior of the building seemed to radiate a pink glow when caught at just the right angle.

A Rivian Bowhead approached, appropriately named after the whale with the largest mouth. The roof was an enormous glass panel that curved and extended down to the front of the car. Instead of side doors, the Bowhead had an enormous cabin made possible by the roof lifting like a whale's mouth, allowing the occupants to exit down a walkway on the front of the car.

Inside, Kasemchai lay across the bench seats of the Bowhead. He rubbed his temples, groaned, and muttered under his breath, "Why do the Labonnes have to visit on a Thursday afternoon? So many things I could be doing today."

The Bowhead drove through the gate and followed the large arc around a fountain of two enormous marble oranda goldfish, their bulbous crowns so large they sagged over their eyes. Defying nature and science, they appeared to leap through the air. The car stopped before the main stairway. It's large glass roof raised, and Kasemchai started down the stairs that folded out from the front. Several attendants rushed to greet him, and Kosal, head of the household, stepped forward. "It's so good to see you, Mr. Kasemchai! Your parents and the Labonnes are on the swan deck."

"Do you have any bags?" a second attendant asked.

"No," Kasemchai blinked several times under the bright sun. "I'm only here for the afternoon."

Kosal led Kasemchai into the home and up the stairs. "Would you like a glass of iced fruit tea? I find it's very rejuvenating this time of day."

"You mean when I'm hungover?"

Kosal looked offended. "No, sir. I know the old wives' tale you speak of. But at your age, it is best to be as refreshed as possible. It keeps your skin taut."

"Thank you, Kosal." He laughed. "I'm just giving you a hard time. Sure, bring me an iced tea."

Kasemchai made his way through the house towards the swan deck. It was covered with a large arching roof and had an incredible view over the gardens and the ocean. A canopy of thick bougainvillea vines with purple flowers cast soft sunspots on the ground. The Rain Wall to the left could clearly be seen—the source of the Seng family's enormous wealth.

As they neared the door to the deck, Kosal added, "The Labonnes have brought someone on this trip. I think you know her."

Kasemchai's head snapped up at Kosal as they entered the warm, sun-splashed deck. The group chatted together on oversized chairs with red silk cushions and ornate teak frames.

His father stood and beamed when he saw Kasemchai. "My financial advisor is here! Kasemchai, come say hi to the Labonnes!"

Mr. Labonne gripped Kasemchai's outstretched hand and vigorously shook it with both hands. "Kasemchai, my, how you've grown since we last saw you! Sophia, please hug this boy before he thinks we have forgotten who he is."

Mrs. Labonne was already rushing toward him. A flurry of bright purple, saffron, and azure colors clashed with her ample tan bosom. She glittered in gold and diamonds that hung from her arms, neck, and ears.

"Philippe is right, my boy! You have grown. What's it been, two years since we last saw you?" Mrs. Labonne said.

Mr. Labonne placed his hand on his shoulder and pointed Kasemchai towards his daughter. While Mrs. Labonne was an explosion of color and joie de vivre, her daughter's face was calm and controlled. She tilted her head slightly to allow her dark red lips to draw a smile that masked her thoughts. Her hair had the same straight bangs, but now it was short in the back and angled down toward her shoulders. The ends were sharp and pointed toward the lacy black choker wrapped around her neck.

"Do you remember Angélique, Kasemchai?" Philippe straightened and puffed out his chest as he said, "She is very busy these days, but she was able to join us to see your family."

Kasemchai walked over to her and shook her hand. "Of course. It's good to see you again, Angélique. It would be hard to forget you."

"I'm happy to see you as well, Kasemchai."

Preap beamed at his son and looked over to Angélique's parents. "I'm surprised Kasemchai could join us today. You know he's the main backer behind the Serendipity PrefX? I understand this is how everyone is meeting their soulmate these days. Very busy."

Angélique's father glanced towards Kasemchai's parents. "He looks so strong, Preap! How does he find the time to exercise with such a busy schedule?" He gave Kasemchai a strong pat on his back, then gave his bicep a good squeeze. "You know, Preap, Angélique has been incredibly successful with my *Look Up to God* virtumersion."

He leaned down to pick up his wine glass from the heavy stone coffee table. "I've decided to put her in charge of opening our new water purification center. It is our goal that all the water we move by ship has already gone through micro-filtration, reverse osmosis, and sent through ultraviolet light chambers." He brought the glass to his lips as if he was finished, and then, just at the moment that someone else could talk, he held it aloft and turned to his wife. "I really don't know how she can motivate so many people to stay on task. She gets it from you, doesn't she, Sophia?"

"Angélique, I'm certain it will be another architectural masterpiece," Preap said, his voice full of admiration as he glanced thoughtfully toward her father. "Much like your father's Crystal Egg—bold and unforgettable."

Kasemchai smiled approvingly with not a hint of the sarcastic thought that went through his head: *How could a water purification plant look like the Crystal Egg?*

"Kasemchai, you must take Angélique to see your office tonight," Preap said, turning to Angélique's parents with a gleam in his eye. "You must have heard about the incredible night view from the AquaSpire. It has transformed the city."

Philippe nodded, his gaze shifting warmly to his daughter. "Of course, Preap. If anyone has an eye for architecture, it's my Angélique."

Kasemchai leaned back with a relaxed smile, catching Angélique's gaze. "Well," he said loud enough for everyone to hear, "if our parents tire of talking about us, I'd be more than happy to give you a tour of the building after dinner."

Angélique chuckled. "Will they be finished by then?"

Everyone laughed.

Philippe sat up straight. "It is truly a blessing how our fortunes have grown over these last sixteen years. You know the hardships we've endured as well as I, Preap." He lifted his glass. "To the rain. It is the giver and the taker." He turned to face his wife and then Kasemchai. "May every day be another day to brag about our children's opportunities."

A servant appeared in the doorway and bowed in their direction.

Preap stood and motioned towards the doorway. "Please, everyone, it appears our dinner is ready."

PREAP INVITED Philippe to the adjacent room for an after-dinner digestive. They shared a moment of laughter as Preap poured a crystal bottle of cognac into two tulip shaped glasses, and the two men leaned in close.

Kasemchai grabbed his glass of wine to listen in. When he was closer, he overheard his father say, ". . . cost of bunker fuel is getting too high for our tanker ships. When water is too expensive for the poor, governments get jumpy. We need to schedule a meeting with the Saudis."

Preap stopped when Kasemchai neared. "Two old men always having the same conversation, Kasemchai. You should tell Philippe about your latest investments."

"Sure, Father. Well, of course, Serendipity is one of the top three dating PrefXs in the world. I can't share any details, but I was in talks with one of the largest AISLs today."

Philippe's brow furrowed, and he tilted his head slightly, eyes narrowing as he wrestled with the acronym. "I never know what to say. Is it pronounced *aisle*?"

Preap dropped his hand on his friend's shoulder and, with authority in his voice, said, "No, I understand they pronounce it *ais-el*."

Kasemchai laughed and sipped his wine. "I don't think it really matters. Regardless, A-I-S-L is an acronym for AI Service Layer."

Philippe chuckled and raised his glass to toast Preap. "You can't expect us old men to remember all this techno jargon, Kasemchai. I just tell my companion to do what I want, and it does it. I don't need to know where it is." He lowered his glass. "This AI Service thing. Are they getting into the dating business?"

Kasemchai smiled mischievously. "They are, but not in a way that you might think. If this goes through, it will be very lucrative."

"I see you get your eye for finance from your father, Kasemchai. It's good to know our family members can make money as well!" Philippe leaned in and grinned at Preap.

"That reminds me," Preap said. "You should take Angélique to see the AquaSpire now. The nighttime view is spectacular. And have Nisay drive you. Our most precious cargo in one car. I want to make sure you're well looked after."

"Of course, Father. I'd be happy to show her around. I overheard you two talking about the cost of bunker fuel. I believe that . . ."

"I'm so glad you are taking an interest in our family business, Kasemchai. This is business for Philippe and I." Preap leaned backward and patted his son's shoulder. He beamed with pride.

To Kasemchai, it was the most degrading look his father could give him.

Chapter Eight

ANGÉLIQUE AND KASEMCHAI sat in the back of the sedan as they drove downtown toward the AquaSpire. At first, the buildings were rounded, with faces that resembled oval honeycomb. They were covered with vines and sponges. Closer to the city's center, they became taller and more angular.

"Why is Sihanoukville the center of the universe for your family?" asked Angélique.

"Well, it's good luck, really. My father worked on a ship, and we were fortunate the port was spared when the wall of rain came." Kasemchai topped off Angélique's glass of wine. "If I really think about it, I suppose it's no different than your father and his Crystal Egg. I think they just want to be where the water is. They need to see the wall."

"Hmm, I suppose you're right." Angélique leaned toward the window to see the tops of the buildings and the hanging vines. Each block into the center became more crowded, and the sidewalks were filled with people. "It seems like there are twice as many tall buildings than there were two years ago," she said. Then, after a pause, "So green."

"Of course. The jungle never stops growing, and neither does our city."

She rolled down the window and let the rain fall on her outstretched hand. Her head tilted further as she spied a pack of girls on the corner. Translucent gel covered their heads and dripped down or swooped over one eye. Their glossy lipstick matched the different gel colors.

"What's up with those girls?"

Kasemchai leaned toward the window. He grinned. "They call them Jelly Girls. They're supposed to look like they're emerging from the water. Something like that. "

Angélique giggled. "I don't think I could pull it off, but to each their own."

The sedan arrived at the AquaSpire. They entered and rode the elevator up to the fifty-sixth floor, where doors opened to a brilliantly white room. Kasemchai led her to the end and then leaned forward to open both doors at once—a favorite flourish of his to show off the grandeur of his office.

When Angélique entered, she made a low whistle as she took in the two-story ceiling. At the far end were tall windows that looked out onto a lush garden. And beyond that was the ocean. The sun was setting, and it was a brilliant wash of colors from saffron to violet.

Kasemchai turned to close the doors. They clicked shut, and then he yelped.

Angélique pressed against his back. Her hand gripped his thigh as she leaned close and nibbled on the tip of his ear. "Really, Kasemchai? 'It would be hard to forget you.' Don't you think that's a little obvious?" Her breath was warm and sweet.

Kasemchai turned and pulled Angélique into him. His mind raced back to their last encounter in a small room on the Ivory Coast. Then, he was back in the present. They kissed aggressively as if gasping for air. He could feel her teeth bruising his lips.

Moments later, they eased apart, still holding onto each other as they leaned against the cool column by the window, calming their racing pulses.

Angélique looked out the window at the garden. "I'm going onto the terrace," she murmured, gracefully adjusting her dress. She paused, glancing over her shoulder with a playful tilt of her head. "Are you coming?"

"Of course. Let me grab some water."

The massive glass door moved effortlessly when she pushed it open and stepped into the warm, moist air. Outside, the ceiling was two stories high. Vines draped down like curtains and swung gently in the breeze. Enormous birds of paradise reached up, bathed in soft light. She could almost hear the traffic below, but a soft rain dulled the sound. She sat on one of the orange cushions and curled her feet underneath herself.

Kasemchai brought a carafe filled with water and ice cubes and placed two tall, thin glasses on the table. He added a wedge of lime to each glass and poured water into them. He then handed one glass to Angélique. She took the glass and rolled it over her forehead to cool down.

He flopped onto the couch and slunk down further. His right hand was on her lap, and his left held his glass aloft, toasting the ocean. "You should visit more often, Angélique."

"You don't think every two years is enough?"

"Of course not!" Kasemchai claimed with a wounded look. "I've been completely celibate since I last saw you. I can't possibly go two more years."

"Sure you have, Kasemchai. And I've been pining away for you this whole time. Managing the *Look Up to God* virtumersion. Breaking ground on a new water purification center."

He looked at her earnestly, rubbing his hand lazily up and down her smooth skin. "You have been busy. And it sounds really meaningful, Angélique. You're building real things. My business is growing well, but it's easy to be a VC with other people's money."

"You're not being fair," Angélique protested. "Serendipity is a huge success. You're bringing lovers together."

"Sure, I read the endorsements. The heartwarming, life-changing stories. But it's a matchmaking PrefX. It's not like the world really needs another matchmaking PrefX."

"Well, what do you want to do?"

"I don't know. I have a new thing I need to focus on. It's complicated."

"Oh, I see," Angélique mocked. "It's a complicated thing."

"Who am I kidding? Who wouldn't invest in a Seng's business? I'd rather be building something of substance, like your water purification center. That's a tangible thing." He stared out towards the ocean. "Something my father would actually be proud of."

"What do you mean? I know your father's proud of you. He couldn't stop talking about you tonight."

"That's just because I'm his project," he said, feeling the weight of the words. "Just a tool to parade around for admiration. He wants to show your parents how good his son is." Kasemchai scowled at his glass, then jerked his head up. "After dinner, I joined our fathers to hear what they were talking about—fuel for their tanker ships. But when I approached them, they stopped."

"Poor baby." Angélique pouted. "To rule your world is not enough. You must rule our fathers' world as well." Then, she pressed her finger against his lips. "You need to get away. There is an island called Ishigaki Jima. It's far enough from here that you won't see anyone you know. Do you know why it's famous?"

"No."

"It's one of only two places where you can get Japanese black pearls."

"And you want me to bring you a string of black pearls back." A slow smile crept across his face when he recognized her obvious ploy to buy her jewelry. "I see now why you're sending me on this quest."

"No. But yes, you do need to buy me a string of black pearls. It's a tribute to me. Or maybe—" She repositioned herself on the couch, sitting up straighter. "I'll join you." She waited a moment but he didn't catch on to her offer, so she continued. "They have giant manta rays. Once a year, the giant manta rays return to the island. They come for mounds of coral beneath the sea where tiny fish gather." She placed the back of her hand on his thigh, dragged the tips of her nails upward, and asked, "You're a scuba diver, right?"

"Of course."

"I thought so." Angélique stared deeply into his eyes. Studying them, perhaps trying to see through them. "It's a kind of health spa for giant manta rays. As they fly by the coral hills, cleaning fish latch on for a full body exfoliation treatment. You dive down to these mounds, hold on to the coral, and wait. The manta rays will swim directly over your head if you are still. It is, I hear . . ." she waved her hand over her head, "spiritual."

Kasemchai leaned forward. He gently pulled her head towards him, their mouths so close they could kiss. "Very well, Angélique. Let's go see these giant manta rays and find the largest string of black pearls for your neck." He traced her cheek and the curve of her jaw with his lips, then grabbed the lace choker in his teeth and tugged.

Angélique sat back and giggled with delight. She stood up and walked over to the railing, hurrying as she neared. She gasped. Kasemchai smiled and approached behind her.

"What is it?" asked Angélique.

"Did you think the lights my father spoke of were the city lights?"

They looked over the railing down to the ocean. Each raindrop exploded in light as it ricocheted on the water's surface. A soft trail of blue lights could be seen along the beach, outlining the tension between the sand and the sea.

"It's bioluminescence," Kasemchai said. "The waves and each raindrop disturb tiny creatures. They respond with light—millions of them. And for us, it's a wonderful show."

"Incredible, Kasemchai. I'll think of this when I'm working on the purification plant."

Kasemchai stared out at the ocean. "You should join me in Ishigaki Jima."

Angélique looked up at him. "I'd love to."

"Are you certified for scuba?"

The blissful smile on her face dropped ever so slightly. "No."

"You should get certified before we go. It wouldn't be worth the trip if you can't dive with me." He leaned over and kissed her forehead.

Chapter Nine

THE NEWARK WINERY was a six-story building with a large one-story warehouse adjacent to the dry Passaic River. The outside of the building was covered in a thick coat of dust. Three years earlier, it rained for a magical ten minutes, leaving grimy streaks down the sides of all the buildings in the area. Inside the building was a single cavernous room. It was pristine.

Makoa walked between two of the towering racks that filled the room. On the shelves were grapevines. A thick trunk grew nearly two feet high and then out in two horizontal branches along a wire. Shoots grew vertically along the branches from which the grapes could grow. It looked like an absurdly wide pitchfork. This continued in a row from one end of the building to the other.

LED growing lights were placed over the shelves of grapevines, and the stacks of shelves repeated over and over up to the sixth floor—an entire vineyard of gnarled branches and lush green leaves stacked vertically in twenty-four towers.

He stopped to inspect a dark red cluster of grapes hanging nearby. His hand reached out toward them, then dropped to pick up a pinch of the granules from which the vine stalk grew. He smelled it, tasted it with his tongue and smiled before dropping it back into the tray.

The granules mimicked the Côte d'Or, or "Golden Slope," in France's Burgundy region. This soil was rich in lime, but the chemical secret was based on a specific depth, where the roots actually dove below the base rock to gather their nutrients. This was the exact composition that produced the extraordinary Pinot Noir grown in the Newark Winery.

Makoa stepped back when he heard a subtle beep, an audible nudge from one of the robot arms. They moved up and down the racks, suspended on two thick wires that allowed them to go in any direction on a horizontal plane. They carefully picked up each grape cluster to evaluate its dark red color, firmness, sugar content, pH, and titratable acid levels. If a cluster was at its optimal level, it was plucked and placed in a bucket on the right. If it was damaged, it was placed in a bucket on the left side, along with dry twigs and leaves the robot had scavenged.

Makoa watched the robot arm for a moment, then turned and headed for the exit. He passed an area filled with a cacophony of stainless steel machines for measuring. Just after exiting the door, he stopped to splash his boots in the tray of disinfecting solution.

In the break room, he sank down into one of the chairs surrounding a low table, placed a small bowl on it, and tugged off the plastic lid. It was filled with salted crickets. He dropped a few into his mouth and then placed the disc from his AugWatch next to the plastic lid.

"Hey, Gene. Play the latest *Look Up to God*."

His companion said, "Sure thing, man."

Makoa sat back and smiled as Philippe's warm face turned towards him.

"Thank you for joining me, my dear friends. As I was walking towards my chapel this morning, I was thinking . . . I was thinking about the haves and the have-nots. Why has God given so much water to some? And so little to others?" Philippe looked down at his closed fist as if the answer might be inside. "I was anguished as I thought of the people who suffer. I wondered if they were being punished." He peered up into the sky, frowning. Then, a large smile crossed his face as a warm light shone upon it. "No, these are dark thoughts. God does not punish us. This is the devil distracting us, trying to trick us. To be in the arid land is not to be forsaken. God has given me water to share with the world."

He stood to pace. "The rain still falls from heaven. That is the only true water. In a world where we turned our backs on our neighbors, where we thought only about our needs, with the bounty he shared daily, God tested us with the drought and the Rain Wall. The drought, the rain, the flood—are these not stories directly from the Bible? You've heard of them before. I've shared these stories directly with you, my friends.

"Yes, he tested you who live in the drought. Then, he shared his lesson. He blessed us with a path to share our water from the flood lands. He helped us see that we are all on one earth, that we need to cooperate, and that brothers and sisters can share our bountiful resources with those in need."

He stopped, stared into the camera and frowned. Makoa felt like he was talking to him directly. He stopped chewing the crickets, and his jaw hung slack.

Philippe inhaled deeply through his nostrils. "Don't be fooled, my friends. There is no shortage of water. God showers water on us daily. Don't question God. Don't ask if he's forsaken you. God loves you and thinks only of you. Each and every day, you should cherish his gloriousness. You should accept his offer. Water from God is God."

Makoa's AugWatch vibrated, and Philippe's voice paused just before he began to say his next word.

"Makoa, are you free?" The winery owner spoke quickly, as he always did. "We're getting ready to rack the chardonnay."

"Yep, I'll be right there," Makoa said. He snapped the disc back onto his AugWatch, grabbed the container of dried salted crickets, got up slowly, and started towards the door.

His companion spoke to him. "Back to work, man. You got this."

Makoa enjoyed listening to modern music as much as the next person, but he also loved the classics. His all-time favorite band was Kiss, and he had chosen Gene Simmons as the voice for his companion. Gene was a man of patience and power. He could respect that.

"Thanks, Gene."

"You're welcome. And Makoa . . ."

"Yeah, man?" Makoa heard the bang of a drumstick setting a heavy beat, and then Gene sang "I Love It Loud" at the top of his lungs. Makoa pumped his fists in the air as he headed back into the growing room.

Chapter Ten

KASEMCHAI'S JET screamed across the ocean and touched down on the tarmac at the Ishigaki Jima airport. Three hundred miles south of Okinawa, the name simply described the island's early human modification. In Japanese, "ishigaki" meant "stone wall," and "jima" meant "island."

He looked down at the AugWatch on his wrist. "Chanthavy, is Angélique available?"

"Let me check," Chanthavy said.

A moment later, Angélique's avatar appeared in the jet cabin. It sat down in the chair opposite Kasemchai.

"I'm just landing in Ishigaki now," Kasemchai said. "When are you coming?"

"I'm sorry. When am I coming to Ishigaki?"

"Yeah, I sent you my travel itinerary. I told you I had a house there."

"Yes, but Kasemchai, you never invited me."

Kasemchai cocked his head to the left and stared at the ceiling. "In the AquaSpire, don't you remember?"

"You told me I could come if I became a certified scuba diver. I don't have time for scuba training, and that wasn't an invitation—it was a condition."

Kasemchai's temper was rising. He stared hard at her, then let his temper subside. "You're right. I wasn't clear. I should have confirmed with you."

She pursed her lips and, in a pouty voice, said, "Please don't hate me, Kasemchai. I would have loved to be there with you."

"No, no. Don't worry, Angélique. I'll visit the manta rays, and then I'll find you a pearl necklace. Besides, I wanted to check out some old awamori distilleries while I'm here. Boring stuff."

"Next time, Kasemchai. And I'll be waiting for my necklace." She blew him a kiss, and her avatar disappeared.

Kasemchai stared out the window at the tarmac and watched two Range Rovers approach. A large portable roof was moved toward the jet to keep his side in the shade. When the jet's door opened, the humid heat punched through it. He closed his eyes, took a deep breath, and exhaled slowly. After opening his eyes again, he made his way to the door.

As he walked toward the second Range Rover, an attendant asked him for his drink order. He was feeling dehydrated from the flight but opted instead for a double whiskey over ice. Inside the car, he reclined the seat, sipped from his glass, and sulked.

The two Range Rovers wound their way through the streets toward the ocean. The roads grew increasingly narrow the further away from the airport they got, often lined on both sides by the island's namesake stone walls. Kasemchai kept himself entertained by scanning the orange terracotta-style roofs, looking for the shîsâ, a lion-dog guardian people placed on their roofs to guard against danger.

At last, they arrived under the shade of the entrance to the house he had reserved. When the door was opened, he stepped out of the car and was greeted by a line of attendants on either side of the front door. They clasped their hands in front of themselves and bowed.

In unison, they called out, "Irasshaimase."

One man stood in the front. He bowed carefully so that he wouldn't spill the tall, thin glass of champagne he held. He hurried forward to hand it to Kasemchai. "Welcome, Mr. Seng. My name is Kato. I will be your attendant during your stay. I hope your flight was smooth."

He led Kasemchai into the foyer. Beyond the lounge chairs was a large wave-shaped wall with gold-leaf ginkgo leaves on a cream background. The entire room felt very tranquil. When they entered Kasemchai's bedroom, he realized the building was perched on a cliff and overlooked the sea.

As he walked towards the large windows, several maids stood off to the side and bowed in unison. Kasemchai picked up a flower from the vase on a table and handed it to one of the maids.

"Thank you," he said. Then, with his champagne glass in hand, he gazed over the blue ocean.

Kato walked up beside him. "I have made your dinner reservations at Yaeyama for 8 p.m. tonight."

"Thank you, Kato. I've changed my plans. I'm going to eat on my yacht tonight," Kasemchai responded. "Let the captain know. And could you do me a favor?"

"Anything, sir."

"I'd like to sample some local awamori as well."

"An excellent idea. I will have a hostess prepare a selection for you."

Kato moved through the room carefully, his face always trained on Kasemchai while he checked for dust or anything out of place. The queen-size beds had been removed from his suite, and a king-size bed had been flown in per Kasemchai's request.

"Your appointment with your golf pro is tomorrow at nine," Kato said, his gaze focused and attentive. "Does that still work for you?"

"Move it. I'd like to go to the gym at nine. Maybe I'll play in the afternoon."

Kato smiled contentedly as he adjusted the flowers in the vase and said, "Of course, sir. I also spoke to the captain about your dive. The optimal time for diving with the manta rays is between noon and three. Have you dived with giant manta rays before?"

"No, a friend recommended I come here to see them."

"Oh, fantastic. You're in for a real treat. They are so majestic."

"That's what I hear." Kasemchai took a sip of champagne. "Have the car ready. I have an appointment at two."

"Of course, sir." Kato bowed and shooed everyone out of the suite.

Chapter Eleven

LIV LANDED on the tarmac at Painushima Ishigaki Airport. She had already cleared customs in Tokyo and was on a thirty-person jet with Japanese tourists. The plane doors opened, and a wave of hot, humid air blasted through the cabin. She immediately began to sweat as the passengers behind her pushed her forward into the passengers in front of her. The cabin grew hotter and staler with each minute.

Once outside the door, she grabbed the railing and carefully walked down the stairs. They swayed with the rhythm of each passenger repeatedly taking one step, then dragging their bag one step. The black tarmac sizzled, and she followed the crowd toward the main building.

"I suppose this is the way to baggage claim," she said out loud. She secretly hoped someone would answer, but no one did.

Fortunately, she was right. They entered the building, and the crowd walked over to the baggage claim. A loud buzzer sounded, and red lights flashed over the door where the bags would emerge. The carousel slowly clunked by for several moments and then stopped. The buzzer stopped, and the red flashing light stopped. Everyone around her started to murmur, and an announcement was made over the loudspeaker.

She asked a gentleman next to her in English what was happening.

He blinked, his brow furrowing slightly, and then a small, polite smile flickered across his face. He made no other motion.

Liv asked Noa to translate for her. She pointed at his AugWatch and made a motion like a duck quacking, hoping he understood she was asking him to turn on his AugWatch translator.

He pointed to his AugWatch and then flashed five fingers. Then he pointed again and flashed five fingers. Again, he pointed to his watch and flashed five fingers.

Good god, she thought to herself. Did he mean five minutes or fifteen? And why didn't he just answer in Japanese?

Liv smiled and said, "Arigato."

The man bowed. Liv bowed. The man bowed again, more deeply. Liv bowed more deeply. He bowed again. She bowed again.

Liv waited. She had been flying for an entire day. After several hours in Narita Airport, she was standing in a building so humid that she felt like she was standing in a shower. She wondered if her luggage even made it to its final destination.

After twenty-five minutes, the buzzer and the flashing light came back on. A weak cheer lifted from the crowd. The bags started to tumble out the door and onto the carousel. When her bag finally appeared, she grabbed the handle and pulled it. It was enormous, and it wouldn't budge over the lip of the carousel, dragging her along with it.

The man she spoke to earlier rushed over and helped her pull the bag over and onto the floor. Once settled, he bowed, and she bowed. He bowed again. She bowed again. He bowed again, and she bowed again.

"Arigato," she said. Her R sounded more like the R like the name "Ari" than the Japanese "arigato" that rolled the R like "arrivederci," but the man didn't correct her.

"You're welcome."

"So nice," she said out loud. "Wait a second, do you speak English?"

But he had already walked away.

She laughed, pulled the handle out on her luggage, and headed to the exit. She found her way to the shuttle area and discovered a woman holding a sign with her name. She was clearly staring at Liv but made no move toward her.

Liv approached her and blurted out, "I'm Liv. Liv Anselm."

Her face beamed. "Nihongo desuka? Speak Japanese?"

"No, I'm afraid not," answered Liv.

"Ah, so des neh." She drew out the last word as if she needed to finish her thought before she could complete the word. "I speak English."

She took Liv's luggage and steered it toward the back of a rusty old van. After the two women managed to heft the bag into the back together, she opened the van's sliding door and pointed with her open hand for Liv to enter. The door was sandy and creaked, but Liv smiled. It was a good sign that they were soon headed to the beach.

"My name is Haruka," the woman said.

Haruka drove along the coast for an hour to Kabira Beach. She asked Liv if it was her first time to Ishigaki Jima, what food in the Netherlands was like, if they really had windmills, and the obligatory question everyone asked: "When had it last rained?" For people who lived in areas without rain, the conversation always ended up there.

At last, they stopped in front of the dimly lit Bagaa Jima dive shop. It was unimpressive. The beige paint was dull and chipped on the outside, and the blinds were faded. The inside was no different.

Haruka pulled out gear for Liv. She tried on a few of the graphite-colored dive skins until she found a good fit—not too baggy to let in the water but not so tight that she couldn't breathe.

She removed the dive skin and pulled on booties to try on the long dive fins. While she was trying them on, Haruka brought over a tank for her. She buckled on a thin harness and then spun it around and slapped one of the two bumps on the side of the tank.

"Do you know how to use a hydrotank?" Haruka asked.

"Sure. The buoyancy control is inside the tank, no weights or BC."

"That's right. This one is self-regulated, so you don't need to control the pump. As you use more oxygen, the automatic pump will add seawater to make you neutrally buoyant."

Liv ran her hand over one of the two small cylinders near the top of the tank. "And when I get back up to the surface, how do I dump all the water?"

"It's a smart system, Liv. It dumps all the water for you."

"Amazing."

Haruka went back to the rack of gear and rummaged in a box. She pulled out a long, narrow gadget. It was black and yellow with several covered propellers and a large camera lens on the front. "This is your underwater drone. I'll sync it to your AugWatch, and it will follow you underwater. You'll have an incredible virtumersion to take home."

After that, they locked up and hopped back into the van.

VEERING DANGEROUSLY close to the stone wall on the right, Haruka swung left and into the driveway of the ryokan—the hotel where Liv would stay for the week. The two women jumped out and wrestled the heavy luggage to the street.

"Livu-san, follow me," Haruka said. She helped Liv drag her bag up the stairs to the door and introduced her to Mr. Nakamura.

"Livu-san, welcome to Ohama-so Ryokan!" Mr. Nakamura shouted warmly. "Please, please, follow me inside."

"Wait. Haruka, how will I get to the dive center?" asked Liv.

Haruka had already started down the steps. She turned and flashed Liv a reassuring grin. "Don't worry, Livu-san. Later tomorrow, I will take you to the dive boat."

"Later?"

"Later. After lunch," responded Haruka.

"Got it. Arrogato."

Haruka and Mr. Nakamura laughed at her mispronunciation.

"Your Japanese is very good!" Haruka said.

"Yes, Livu-san, like you were born here," confirmed Mr. Nakamura.

Inside, Liv was greeted by a large woman who didn't share the same smile as Mr. Nakamura. She was no-nonsense and booked Liv into the hotel. She checked her passport, looked at it disapprovingly, and then called out to Mr. Nakamura, who reappeared from around the corner.

He showed Liv around the hotel and up to her room on the second floor. It was long and narrow, with a faux tatami floor. On one end of the room, three thin mattresses were folded up in thirds. A pillow was placed on top of each one.

Liv rolled her bag into the room and placed it on the floor next to one of the mattresses.

"Livu-san, you can settle in." He pointed to the left. "The shower is down the hall." Then he turned and pointed in the other direction. "And the beach is in that direction. We will have dinner at six."

Liv thanked Mr. Nakamura and looked around her room. Apart from the three mats, the room was bare of furniture.

"Looks like I'm living out of my bag this week," Liv said to herself.

After taking a shower and changing into fresh clothes, she realized there were still several hours left until dinner and decided to go find the beach.

Chapter Twelve

Kasemchai's digital companion recommended he visit a local distillery during his trip to Ishigaki Jima. A big fan of sake, he wanted to try awamori, a traditional beverage from Okinawa.

The building was aged with wooden walls that let the sun shine through gaps in ghostly rays. Kasemchai's guide explained to him that Mr. Tanaka did not have an AugWatch, so she would have to translate for him. Inside was cramped, and only the guide and Kasemchai's attendant followed him into the building; the driver and bodyguards stayed outside with the car.

The virtuoso distiller, Mr. Tanaka, showed Kasemchai the ceramic pots where the rice liquid was fermented. He made a sprinkling motion with his hand, and the guide explained how black koji spores were sprinkled over the rice. Sake used yellow koji spores, but because of the hot, humid climate, they needed to use black for awamori.

Then, Mr. Tanaka scooped up the warm rice with his hands and held it up for Kasemchai to smell. The steam washed over his face, and Kasemchai smiled.

"Mmm, it smells delicious!"

He spoke to his guide, who asked the distiller how long the liquid was stored in the pots.

"Three weeks," Mr. Tanaka said. "But that is only the fermentation process. From here"—he pointed to the taller clay pots—"we store the awamori for a minimum of three years."

With his gnarled fingers, Mr. Tanaka motioned for Kasemchai to follow him.

"Three years is good awamori, but he's going to show you his special collection," the guide said.

Mr. Tanaka opened a wooden box. It was filled with bottles made of a silver metallic ceramic. Each bottle had a sharp-edged dragon carved on the side and cloth wrapped over the top. He picked up one bottle and brushed off the top.

"He said he brought this out from his cave when he heard you were coming," the guide said. He had been trying to look professional, but his eyes widened when he looked over the bottles. "These bottles are fourteen years old."

Mr. Tanaka carefully unwrapped the cloth from one of the bottles. It was painstakingly slow, and his hands trembled. Kasemchai resisted the urge to help. Besides, the wait made it seem that much more precious.

Kasemchai sat on a stool around a low table and enjoyed a drink with Mr. Tanaka. "Mr. Tanaka, it's amazing. I'd like to own this distillery."

His guide translated. Her face flushed as she relayed Mr. Tanaka's message. "I'm sorry, Mr. Tanaka would like to know why you want to own his distillery. He said you can drink his awamori anytime you wish. It would cost far less."

Kasemchai looked at Mr. Tanaka and said, " Because I can, Mr. Tanaka. Cost is not the point. I wish to own it. It's so delicious. I want it all."

She again delivered Mr. Tanaka's message. "But awamori is for drinking with friends. If you were to own his distillery, and, I'm sorry, he said you can't, then you would deny so many people the opportunity to drink it."

"I'm comfortable with that," Kasemchai said. He motioned to his assistant. "Please, take care of the details."

He stood up and hurried towards his car. Outside, his movement slowed to a stop when he spied a woman walking down the middle of the narrow road. Her long blond hair was striking on an island where every person had black hair. Still wet, it dripped down on the top of her fuchsia bikini top.

She turned towards him and smiled. A broad grin spread across his face until he realized she was not looking at him. She looked *through* him as if in a trance. When she turned back towards the sea, it was as if she were staring at a newborn baby. Her lips curled up in an impish grin.

Kasemchai felt drawn towards her. He could almost feel her eyelashes brush against his. His gaze traversed down and stopped on the thinnest two strands of gold chain that wrapped around her waist, teasing her belly button. She held a short stick in her hand and absent-mindedly brushed it against her thigh. He loathed that stick. Hated it. If only he could be that stick, caressed by her soft fingers.

His hand reached out. He winced when he touched the hot side of the car. He had not been moving towards her. His feet were rooted to the ground. A burst of energy came to him, and he rushed back inside and pulled his assistant outside with him.

"Worry about the distillery later. I need you to find out who that woman is down the street. Do you see her?" He pointed excitedly. He tried to be nonchalant but failed miserably. "Find out what she's doing tomorrow. I want to meet her."

Chapter Thirteen

AFTER DINNER, the guests wandered into Ohama-so Ryokan's lobby. It felt more like a living room with several different styles of couches surrounded by two coffee tables of unmatched heights. On the walls were old faded pictures of divers, and kitschy art made from shells and woven grasses. The windows were covered with bamboo curtains that sagged under their weight.

Liv met several of the guests: a paramedic and two college boys from Osaka, two women, and a mother with her nine-year-old daughter from Tokyo.

Mr. Nakamura never tired of his role as master of ceremonies. His face was creased from a lifetime of sun and e-cigarettes. The long purple vape never left his knobby fingers except for when he lifted up a shiny sanshin to play a song. The sanshin was an Okinawa guitar that resembled a small banjo with three strings. The top head had three long knobs sticking out, and the snakeskin-covered body was square.

He pulled out a second sanshin from an old wooden box and motioned Liv over for lessons. Within half an hour, she had learned several chords and was playing something that strongly resembled a song. When practicing, she screwed up her face—lost in concentration. At last, she placed her flat hand on the strings to pause the notes to look up, and the room erupted in laughter and clapping. The smile on her face was broader than it had been in years.

A few more guests wandered into the lobby, and Mr. Nakamura jumped up and immediately grabbed everyone by the hand. He positioned them into two parallel lines, turned up the music, and gave everyone an impromptu dance lesson.

Okinawa-style dancing was an adult version of the hokey pokey. It was impossible not to laugh as people put their left foot forward and brought them back, then put out their left foot again and waved their hands in the air. With the purple vape firmly in his mouth, Mr. Nakamura held his hands over his head and spun around while clapping. Liv felt incredibly silly but loved every moment. It felt good to not think about work or water.

Panting, they returned to the lobby and fell onto the mismatched couches, their cheeks flushed from dancing. A low hum of conversation mixed with the salty air from the ocean. Shortly after, the mother got up to leave, and her daughter, arms crossed and brow furrowed, begrudgingly followed her. The rest settled in, pouring small glasses of awamori and drinking them eagerly, their laughter gradually filling the room. They shared stories about life in Japan and the Netherlands, hands gesturing animatedly as they described travels and the small quirks of life. Mr. Nakamura quietly strummed his sanshin, and when the conversation died down, the chorus of cicadas outside the windows pulsed in the night.

Liv hardly remembered lying down to sleep on the tatami when she was up and eating breakfast: grilled mackerel, dried seaweed, and tamago kake gohan, a bowl of rice with a raw egg.

The two women from Tokyo showed Liv how to scoop up the egg and rice with her seaweed. She gagged a little as she swallowed the raw egg. Her face couldn't hide her displeasure, and this made everyone laugh.

Haruka parked the van outside the ryokan, and Liv, the two students, the paramedic, and the women from Tokyo all jumped in. It was a quiet ride through the narrow streets bordered with stone walls. They parked beside an opening in the jungle and sleepily exited the van to trudge through a tunnel of leaves until they stepped out into the sun again. Liv caught her breath as she took in the shining white sand that spread out towards the ocean.

The sand contrasted sharply with the emerald green of the ocean and the clear blue sky. The bay was protected from the ocean by a string of black and bleached gray jagged rock islands. Their surface looked like lava with its sharp, porous face, but it was actually coral deposits pushed up eons ago by the Earth's movement. Dive boats lined the shore, appearing to hover over the crystal-clear water.

Liv followed Haruka, and they passed by several large boats. The crew stood on deck like pirates in search of new sailors. Her gaze was drawn to the tall windows, imagining a refreshing drink in the air-conditioned cabins on the journey to the dive site. At the far end of the beach, Haruka gestured toward a boat and said simply, "Here."

Liv realized that her boat was significantly smaller. The flat-bottom speed boat was no more than sixteen feet. The stern featured a small captain's bench and windshield, with a simple roof made of a tarp strung across poles strapped to the railing.

Air tanks were lined up in a row alongside a stack of long diver's fins. Her captain was a jolly-looking man with the tan and ruddy cheeks of someone who spent every day on the water. He helped each person climb onboard and started to explain when to gear up.

Mid-sentence, he stopped, looked at Liv, and pointed at the AugWatch on her wrist. "Liv, AugWatch?"

She looked blankly for a second and realized she should be using her AugWatch translator. She asked Noa to begin translating.

As the captain spoke, she heard Noa's voice translating every word of his into Dutch. It was slightly odd because the captain's lips didn't always match what Noa was saying. Even with the translator, the captain included many pantomimes for Liv's benefit.

Liv was focused on her dive mask when she sensed everyone was looking out to sea. A brilliant green boat sped towards the beach, and a spray of white sea flew out to the far sides. She stared, not sure why this boat appeared so peculiar.

It throttled back and dropped down into the sea. That was when she realized it was on foils. But she soon learned that wasn't its most unusual feature. As it neared the beach, it crawled right out of the water, carried by two large tractor treads that had emerged from the hull.

She turned to the captain in disbelief. "Am I dreaming?"

The captain laughed. He didn't need a translator to know what she was saying.

The boat finally came to a stop. A man stood, walked to the stern, and jumped onto the beach, his skin glistening in the warm sun. Liv stared at his well-formed abs. Then, when she realized he was striding right up to them, she turned away.

He shouted in English, "Permission to come aboard, captain?"

The captain shouted back, "Come aboard, sir!"

Kasemchai looked at the group and said, "Hello, I am Kasemchai."

Liv realized at that moment that the taut abs belonged to one of the most recognizable faces in the world—the only son of the Seng family, the second-most powerful water family in the world. Lead VC behind the Serendipity PrefX. And the ruggedly handsome face often seen below the headline, "Most eligible bachelor."

He stared directly at Liv. "Is it all right if I speak in English?"

Liv's mouth hung open. She caught herself and closed her mouth. "Of course. I'm English." Her face flushed when she realized her blunder. "I'm sorry, I'm from the Netherlands. We pretty much all speak English, though."

"Fantastic!" Kasemchai said. "Are you aware we have the finest dive guide in all of Ishigaki?"

"I . . ." Liv looked around, confused, not sure what to make of what was happening. "I just showed up here."

"Well, not me." A wave rocked the boat and Kasemchai stepped forward, slightly off balance. Liv reached out to steady him and touched his knee, then drew it back as if she'd been shocked. Kasemchai chuckled. "Woah. As I was saying, I wanted to visit the giant manta rays. When I asked who should take me out to see them, it led me to the captain of this boat."

Kasemchai introduced himself to each person on deck. His staff carried his dive gear over and returned to the green boat. After a little nervous banter, the divers started to pull on their glittery grey dive skins. Soon, they were all hot and sweating. They peeled the top part off and let it hang around their waists. They laughed and begged to leave soon before they died of heat exhaustion.

Liv sat on the narrow gunwale of the boat, sandwiched between the paramedic and one of the women from Tokyo. Before they shipped off, the captain said a short blessing to protect them from evil spirits. As he did, he took a pinch of salt from a small plastic bag and tossed it over each person's shoulder.

Haruka pushed the boat off the sand and hopped aboard. They were off.

As the boat sped out to the open ocean, Haruka explained that their first dive was to a large reef bowl. They hoped to see staghorn coral and brain coral, as well as sea fans and sea whips. Swimming amongst the coral, they would most likely see triggerfish, clownfish, and lionfish. She began to warn them not to get close to the lionfish because of their poisonous spines, but she was competing with the wind and the whine of the engine. She moved to the stern of the boat to sit behind the captain.

KASEMCHAI SAT low in the boat as it sped toward their first dive spot. The sea was fairly flat, but the flat-bottomed boat lifted easily on each swell and slammed down. The captain would throttle back when the bow was pointing too high to spare their backs and butts.

It was impossible to talk over the wind and each diver stared off into the distance, lost in their thoughts. Kasemchai stole glances of Liv. He had not spoken more than a few words to her this morning, but already, he felt closer to her.

It occurred to him that she must have spent a considerable amount of her savings to fly from the Netherlands to this remote island to dive. How brave she was to have done this completely on her own. His imagination wandered, and he imagined they were explorers speeding toward an island with a secret treasure.

Then, the whistling in his ear stopped. Everyone in the boat lurched forward as the captain throttled back the boat. They were at the first dive site.

The boat gently rocked from side to side. People half crouched, trying to stretch their legs but holding on to the railing to keep from falling over. They rubbed their wind-blown faces. Haruka and the captain distributed the tanks and fins.

The day felt cooler when they were flying over the water. Only minutes after they stopped, the heat returned, and they reluctantly pulled on their dive skins and zipped up the fronts, anxious to get into the water.

When Haruka passed around the tanks, Kasemchai noticed how strong her arms were. They were not like those of a bodybuilder, but they had definition. She looked like a woman who could lead a dive trip by day and pull on an evening gown for a night out.

Then he stopped his runaway mind and turned back to Liv. Her silver dive skin made her look otherworldly.

The captain commanded everyone to put on their masks and check their regulators to ensure the air was flowing properly. He sat on the gunwale, his back to the sea, and tapped it with a loud thwack. In English, for Liv and Kasemchai's benefit, he said, "I want back door entry."

Liv giggled.

"What's so funny, Liv?" asked Kasemchai.

"Um." She patted her bottom. "It's kind of slang for . . ." Her face turned a shade of red. "You know, it's a kind of sex."

The boat filled with laughter. Liv, already a little red from the heat, turned an even brighter red, which made everyone laugh more.

"Aha!" Kasemchai said. "I think I understand. What should we say instead?"

Liv thought about it and finally decided, "How about 'backroll entry'?"

Everyone laughed.

The captain shouted, "Everyone, backroll entry!"

Haruka sat down next to Kasemchai. "I'll be your buddy on this first dive, Kasemchai."

Kasemchai had a twinge of angst as he watched Liv grab a hold of her underwater drone and roll backward into the sea. It was too hot to delay.

He smiled at Haruka and held his mask with his right hand and his drone with his left hand. He rolled into the ocean. The sea crashed and foamed around him, and then slowly, the bubbles subsided. Silence. He rolled over, breathed out, and slowly descended. From far away, he could only recognize mottled grey, green, and blue.

A floor filled with coral slowly came into focus. Five more feet, he held his nose to release the pressure from his ears. The coral bed was closer, and he could start to make out anemone tentacles waving back at him. Hundreds of fish moved, gently pushed forward by the current, then back, then forward again. Orange and white clownfish darted into the anemone tentacles and shot out just as quickly.

Kasemchai was mesmerized. He hovered over the coral and now and then gave a quick kick to propel forward. Everywhere he looked, he was blasted with colors and shapes. He grabbed a large piece of pink coral, looking very much like deer antlers, and pulled himself closer. His mask nearly touched the coral as he examined the delicate nubs on the branches.

Then, his face flushed, and prickles ran up and down his body. A sea snake, pewter-colored with black bands, emerged from under the coral.

He froze. His breath caught in his lungs. His heart pounded so loudly that he wondered if the snake could hear it as it encircled his arm.

A second, or maybe fifteen minutes, passed. His mouth was slack, and for a moment, he feared the regulator might fall from his mouth. He watched the two black eyes as the snake carefully swung its head from side to side, examining the intruder. Just as suddenly, the snake looked down and slithered—swam—down under the coral.

Kasemchai gingerly pulled his arm back. Balancing precariously over the coral, he stuck the tip of his fin into the sand. As the tip started to bend, he gave a careful push, propelling his body away from the coral. His whole body shivered compulsively. He kicked harder. He wanted to look back and see if it followed him. The muscles in his legs began to throb, but he wouldn't stop kicking as he pushed forward.

He stopped to catch his breath, and his whole body shivered again. He slowly spun in a vertical circle, wondering what happened to everyone else. The ocean around him was immense, and from the great blue, he recognized someone was swimming towards him.

It was Haruka. She came up closer and gave him the okay signal with her hand. He gave her the signal back. It was a relief to see another human being. Perhaps he wouldn't die today.

Haruka peered at his dive computer. Then, with her finger, she pointed at the gauge. He smiled. She pointed again. He gave her the okay signal. Finally, he recognized she wasn't letting the gauge go.

He glanced downwards and realized he was running low on air, as indicated by a red bar. Only now did he realize how much harder it was to draw each breath. His body stiffened, but when he looked up, her reassuring smile instantly calmed him down. She held her regulator out to him, and he eagerly accepted it. He took three deep breaths and returned it back to her.

The two floated in weightlessness. The ground below was equally far as the surface above them. Time slowed down. He and Haruka shared the same air. Their bodies bumped into each other; they were so close. He didn't care about the air. He felt incredibly comfortable, and he wanted to stay there forever, passing the regulator back and forth. A kiss that wasn't a kiss.

He had made a ridiculous ruse to put himself on this boat to be with Liv. He had not counted on Haruka being equally beautiful. Was it ungrateful to ignore a woman who was literally saving his life at this very moment?

He stared into her eyes. His leg brushed against hers as they slowly kicked to keep from sinking. He watched the bubbles escape her mouth as she returned the regulator back to him.

Haruka pointed up. He needed to go up, but it wasn't what he wanted. Reluctantly, he kicked his fins, and they rose. When the two broke through the tension of the surface, he could hear the pump on his tank expelling the ocean water that kept him down. Tiny waves lapped against his face.

He rose a little higher in the water and removed his regulator. "Thank you."

"Of course," responded Haruka. She smiled. It was a warm smile. Her eyes lingered.

"I saw a snake down there," Kasemchai said. "It wrapped around my arm. That's why I used up so much air."

Haruka held onto Kasemchai's bicep, so close to the surface of the water it splashed against their faces. "What did it look like?"

"It had black and white stripes."

She gripped his arm tightly. "Oh, Erabu umi hebi."

"Is it poisonous?"

"Yes, very poisonous. What did you do?"

He felt immensely grateful to be alive. But it was only luck that he didn't suffer a painful death at the bottom of the ocean. The words tumbled from his mouth. "I was terrified. I could see it opening its mouth wide. I imagined it biting me. Every part of my body wanted to swim away as fast as I could."

"Why didn't you?"

"I don't know. My brain told me to swim away. But some other part of me told me to keep my hand still. I didn't know if the snake was just exploring or working its way up to bite me in the face."

"Usually, we run away to save ourselves. Or attack. You made a third choice. That's the hardest choice."

"I usually know what to do," Kasemchai said between heavy breaths. His heart was racing from the memory of the snake, the shared breaths with Haruka and kicking to stay high in the water so he could talk. "This time, I didn't. Who would have known it would save my life."

Haruka rolled over onto her back. "Fight or flight. And you found a third path. I think you've earned your lunch!"

She began to kick towards the boat. Kasemchai followed. He laughed as he kicked as hard as he possibly could, shaking out the nervous energy.

On the boat, everyone peeled their dive skins down to their waists and stowed their gear. Haruka passed out plastic lunch boxes. Each box contained rice, seasoned mushrooms, and crunchy strips of bright green seaweed. The group ate so ravenously that it was a struggle to keep the food in their mouths, stopping only to share stories of the wonderful world beneath the boat and to laugh.

After thirty minutes, the captain stood up. "Clean up your food, everyone. We're heading to Manta Scramble."

The boat skittered across the ocean. The occasional swells lifted everyone off their seats. They hung weightless above the boat, and then, the boat would drop seconds before they came down. They giggled and shrieked as they slammed onto the hard seats again and again.

After twenty or thirty minutes, the captain throttled back. The sea was calm and indistinguishable in every direction as far as the eye could see. It was only a navigation point. The whooshing sound in their ears stopped, and it was quiet—only the sound of the waves kissing the side of the boat. The air smelled salty and fresh.

Haruka went to the bow to look over the side. She directed the captain until they found just the right spot. The group suited up again and rolled over the side and back into the sea.

When Kasemchai rolled over, he couldn't see the sea floor. They were in a much deeper area with a dark green vastness. He stared, moving his head back and forth, trying to find the bottom. He breathed out slowly when he spied the sandy grey floor. He was falling much faster than he thought he should be. His breath quickened. His mind raced. He knew he wasn't descending so quickly, but the ground was moving toward him.

It was large but not sand, a diamond-shaped piece of putty with two corners pulled out to make wings. The head was distinguishable only by a wide cavity, which was its mouth. Two scoops jutted from either side of the mouth like cupped hands. Behind it trailed a long whip—the tail of a giant manta ray.

His heart raced, but his lungs froze. From its enormous size, he could imagine it was a powerful animal. It shifted indiscriminately from side to side, forward then back, completely at the mercy of the ocean currents. One wingtip rolled down while the other curled up. If a blanket from his bed suddenly came to life and dove into the sea, it would look no different.

He stared, entranced, until he realized he was about to land on the back of this creature. He thrashed his arms uncontrollably in an attempt to roll to the right. He put his arms out, preparing to crash directly on its back. Without making any noticeable movement, the manta ray placed itself two meters away.

He continued to fall to the ocean floor, never taking his eyes off this beautiful animal. The world was moving like molasses, and he cared little about where he placed his hand on the bottom. The sharp coral dug into his knee as he braced himself. Swaying back and forth with the ocean swells, he was anchored to the ground.

The manta was gone. He looked around and saw that he was on the side of a small mound of rocks and coral. All around him, the ground rolled and dipped. The landscape was covered so completely with brittle coral that it looked like a soft, fuzzy carpet. His group was scattered across several mounds. The divers huddled close to each other, looking wildly from side to side. Bubbles rose from the divers on each mound.

He stayed low and stared at the top of the hill. A manta emerged, slowly growing in width like the enormous destroyer from the movie Star Wars. It obliterated the sun as it moved forward. It hovered over him, perhaps enjoying a little tickle from his bubbles. And then it slowly drifted away.

Haruka grabbed his shoulder and pointed to a hill on the right. Two mantas were in tight formation, flying over the hill and down into a valley. He and Haruka pushed off and swam to the edge of the valley for a better view.

Kasemchai stood on top of the coral hill, balancing on the tips of his fins and looking down into the valley to see three mantas. Pivoting to the right, he saw two more. Slowly spinning around, he could see a manta disappear on its way out to sea.

He turned more to the right and stared deep into the mouth of a ray. He could see the delicate ribbing of its throat as it drew towards his face. In a moment of clarity, he realized he was staring into a place he wanted to *see*—not *be*.

Falling backward, he let the manta fly over him. He followed its body down to the ten slits of gills, sucking and bulging as they expelled the water.

When Haruka tapped him to head back up, he was surprised. It seemed like they had only just arrived.

Kasemchai surfaced and dragged his body up onto the boat. He pulled his mask off his face and looked at the rest of the group—nothing but silent smiles.

"Amazing, neh?" asked the captain.

Kasemchai laughed. At that moment, he thought he could give up his whole life to work on this boat and see the mantas every day.

The boat sped back to shore. Each person sat quietly, exhausted but exhilarated. Kasemchai glanced at Liv and gave her a dreamy smile. They were each reliving the moment of seeing the rays.

As they neared the sandy shore, the captain cut the engine, and the boat drifted towards the beach. He tapped Kasemchai's shoulder and pointed to his AugWatch.

"Chanthavy, translate for me," Kasemchai said.

"Tonight, we're going to have traditional Okinawa food and then drink awamori all night," the captain said. "You're all invited!"

Kasemchai's eyes drooped, but he responded enthusiastically, "Count me in!" He looked over to Liv. "The captain was inviting us to eat Okinawa food and then go out for entertainment."

"Oh," Liv said. She appeared just as exhausted. "What is Okinawa food?"

Kasemchai shrugged. "I don't know. Maybe noodles?"

He asked the captain what Okinawa's food was. Kasemchai listened to the female voice of Chanthavy say, "He said Okinawa food, but what they commonly eat here is a kind of pig brain soup, grilled lobster or octopus, and butazeri musubi appetizers."

"What's butazeri musubi?"

"Musubi is butazeri sandwiched between two layers of rice and wrapped in seaweed. It was common to make this with an American food called Spam. The company closed when the rain stopped, so people in this region made their own version. 'Buta' means 'pork,' and 'zeri' means 'jelly.' Would you like to see a picture?"

"No." Kasemchai smiled and looked back to Liv. "He said lobster, soup, and appetizers."

"Excellent!" Liv said. "Count me in then."

Chapter Fourteen

THE GROUP DROVE over to a small nightclub. They sat around three short round tables. At the front of the room was a tiny stage with a microphone that was framed by tall plants. The music was a little loud, the room a little dark. Kasemchai sat between Liv and Haruka.

The captain ordered a bottle of awamori. It was brought to the table in the traditional silver ceramic bottle with dragons carved into the sides, circling the shoulder of the bottle. An ice bucket was placed next to the captain, and he used small tongs to place cubes of ice in every glass.

Liv stared in amazement. "Wow! This is a special occasion. I can't remember the last time I saw real ice cubes."

Kasemchai cocked his head and laughed. "You don't use ice cubes? Why not?"

"Well, I don't know, actually. We have insulated cups. But mostly, we drink quickly so it doesn't get too warm."

The captain shouted out something—the word "ishigaki" was mentioned—and he raised an empty glass of ice.

"Wait, wait, wait," Kasemchai said. He pointed to his AugWatch and asked Chanthavy to start translating again.

The captain repeated himself, "Welcome to Ishigaki Jima! Where we still have ice, and we swim in the ocean every day."

He filled everyone's glass, and Haruka handed them out to the group. They all lifted their glasses to the sky.

"A toast to the ocean"—he looked over to Kasemchai—"and water!"

"To water!" shouted everyone. They all took deep sips of the cool awamori.

Kasemchai raised his glass. "To our captain for showing us the secrets of the giant manta rays!" He swung his glass left and right to regard each person.

"Kampai!" they all shouted and took another deep drink.

"What do you think of the taste?" Kasemchai asked Liv.

"It's good. I've never tried this before."

"It's almost sweet. With the scent of vanilla." He took another sip. "Vanilla, and maybe sesame."

Liv took another sip and looked back at Kasemchai. "I can smell the sesame and the vanilla. I thought it would be very alcoholic, but it's refreshing."

"That's right. And as the ice melts, it transforms. It becomes smoother. You may even notice another flavor."

Liv closed her eyes, took another sip, and opened her eyes again. "So decadent."

Kasemchai treasured the moment, then popped the disc off his AugWatch and placed it on the table. He scrolled through the colorful bubbles, stabbed at one, and up came the pictures from dinner. He stopped on a close-up of the udon soup with pig brain. The pink, squishy pig brain could clearly be seen nestled between a large wheel of daikon radishes and greens.

"I can't believe I ate brain," said Liv. She hugged herself and shivered.

"Do you like brain?" asked Haruka.

"Um, I'm glad I tried it. But I think I like the musubi better. I'm better with butazeri and rice." She looked up at Kasemchai. "How about you? Do you like squishy things?"

"Sure, Liv. It's not uncommon in Cambodia. But I'm with you. I could eat the musubi all day long," he laughed.

Liv held up her wrist and swiped at the dancing bubbles on her AugWatch. "Here's my photos of the mantas," she said.

She used her AugWatch to project the images in the air before them. They found themselves submerged in the sea with the coral beneath them. She waved her hand from right to left to navigate through the photos, and they saw the divers scattered over the manta scramble. In the first photo, Kasemchai lay on his stomach, peeking over the top of a small mound. In the next one, he rose onto his knees and bent backward, parallel to the ground, as the giant manta flew over him, separated by mere inches.

"Incredible!" Kasemchai said. He leaned over and swiped to the next picture. Then the next. He stopped on the photo of him looking directly into the mouth of a manta.

"Wow, can you believe how big they are?" asked Kasemchai.

"I know. They're just so peaceful," responded Liv.

Kasemchai rested his hand on Liv's thigh and swiped to the next photo. He pointed out the two rays flying in unison. "You have to share these photos with me."

"Of course. Noa, share all my photos of today's dive trip with Kasemchai," she said.

"Do you wish to include the photos before you got on the dive boat?" asked Noa.

"Sure, send everything from today."

Kasemchai held up his AugWatch and flicked his hand to share his profile. Liv smiled when she saw his face appear over her device and sent him the photos. She saw that her glass was empty, reached out to the table, and picked up the bottle to top off her glass.

Haruka stopped her. "Sorry, in Japan, you never pour your own glass. A good host never lets your glass become empty."

She picked up the ceramic bottle and cradled it in her left hand, tilting it from the bottom and filling Liv's glass. She smiled up at Kasemchai and filled his glass in the same way.

Kasemchai lifted his glass and clinked with Haruka, then clinked with Liv. He threw his head back and shouted toward the ceiling, "I love this island!"

They all drank thirstily.

Haruka leaned over Kasemchai's lap and rested her elbow and a hand on his thighs. "What do you do in the Netherlands, Liv?"

Liv took a sip of her awamori. "I work for a company that adds solar arrays to tanker ships."

"Uh-huh," Kasemchai said. "What and why?"

Liv laughed. "An array just means a bunch of solar panels working in unison to collect as much energy as possible. And why?" She paused to think. Her head cocked to the right, and she gazed off to the left. "Well, we spend all of our time talking about making more electric cars and jets. What people don't see are the dirty ships that bring us all the clothes and furniture across the ocean."

"And the water you drink," added Kasemchai.

She flushed and held up her hand in protest. "That's right. I'm sorry. I didn't mean to say the ships are dirty. What I mean is that they burn something called bunker fuel. It's literally a sludge from the bottom of oil barrels that's super cheap and dirty. One ship makes as much CO_2 as seventy thousand gas-powered cars."

Kasemchai slapped his forehead. "Oh, I was just going to say how wonderful that is. But when you say it that way, it sounds terrible."

Liv held her hand up defensively. "No, no, no. I'm not saying that's a bad thing. You can't send water everywhere by pipes. We need ships to move the water. In fact, we wouldn't have any water on Ishigaki Jima if it wasn't for the ships that bring it here."

Kasemchai laughed. "Well, now I feel slightly better."

"Good. The thing is, if we could reduce the ships' fuel by twenty or thirty percent, that would be a tremendous cost saving, and the less the engines run, the less bunker fuel they burn."

Haruka sipped from her glass and placed it on the table in front of her. "That's a very noble cause, Liv."

"Thank you," Liv said. "Actually, we're trying to win the Phoenix Global Water Challenge."

"What's that?" asked Haruka.

"It's an annual competition that invests in innovations that can reduce the cost of delivering water. If we win this award, we'd be recognized globally for our technology. Plus, we'd get a huge award."

"Like a trophy?" Haruka asked.

Kasemchai laughed. "No, like money."

"Imagine if we could run the ships completely on solar," Liv said.

"Each ship we convert would be like removing the exhaust of seventy thousand cars."

Haruka sat up straight and slapped her hands on her knees. "So when will we have solar tanker ships, Liv?"

"Well, that's a little harder. The ships already have electric engines that turn the propellers." She raised her voice to cut through the rising noise in the club. "We need to replace the engine that generates electricity. You couldn't get enough solar panels on a ship to replace the engine, and besides, shipping companies don't like cluttering their decks."

"What do you need, then?" asked Kasemchai.

"Well, we can only get this to work on smaller tanker ships. We need a solution that works for something like a Panamax-size vessel. These are ocean-going tanker ships. They're large, but not so large that they can't still squeeze through the Panama Canal. There's lots of technology, and we have to convince the shipping companies. There's no simple solution."

Kasemchai's eyes gleamed with a mischievous spark. "But the answer is so simple. What you need is an investor, Liv. My family wants to save money on oil, but I can't invest simply for the love of the Earth. This has to be business."

"That's just it, Kasemchai." She raised her index finger and stabbed at the air to make her point. "Today, ships slow down to reduce drag and save fuel. We could make ships move faster. That's good business."

"Won't that be difficult to get everyone to change? I mean, ships have been running on oil for decades," Haruka said.

Liv laughed. "You wouldn't know it by looking at me, but we are traitors to convention." Liv raised her glass. "Here's to courageous change!"

Kasemchai and Haruka raised their glasses, and they all toasted.

"To courageous change," they cheered.

The music stopped, and Kasemchai realized a man and woman stood on the stage. The woman was dressed in a dark blue yukata, a type of summer kimono. The yukata had a beautiful pattern of white and pink hibiscus flowers that started at her feet and wrapped around her body. She had tied a white obi with large blue polka dots around her waist and wore her hair in the style of a geisha, sleek and black. Her face was painted white, which provided an excellent background for her bright red lips and the wispy red and blue colors that radiated out from her eyes, not too dissimilar to the tail of a fighting betta fish.

The man also had glossy hair that was tightly pulled back. He was wearing a gunmetal-blue yukata with a stiff obi wrapped around his waist, featuring black horizontal bands. He was cradling a sanshin, its square body covered in the distinctive tan and brown circles of python skin. His hand rested on the strings.

Slowly and carefully, he began to strum the strings of his sanshin. Each note hung in the air before the next. The woman held a large wooden stick, which she used to pound on the drum so slowly it was uncomfortable. It was simple, with just the slightest increase in tempo with each beat. Kasemchai was mesmerized.

The tempo wanted to go faster, but the woman held it back with her stick. Only gradually did she let it build in intensity. Every now and then, she let out a loud howl or yelp. The man then began to sing the lyrics to 'Shima Uta,' a beautiful song that harmonized folk music and rock. It spoke of separation, sacrifice, and love. As the music played, the club became livelier. People talked louder, and some clapped along. Eventually, the music enticed people to get up and dance.

Haruka grabbed Kasemchai's hand and pulled him toward the stage. The captain grabbed Liv and brought her up as well. The table emptied as they all ran up. They held their hands over their heads, clapping and laughing.

Midway through the seventh or eighth song, the woman on stage went over to the wall and opened a huge wooden trunk. She held aloft two rubber masks. People ran to the trunk and rummaged through. When they found the perfect one, they held it in the air like a mighty prize. They laughed, pulled on the masks, and jumped around on stage. The band continued to play.

Kasemchai picked up a lion mask with red curly hair and large white eyes that looked more comical than ferocious. As some people jumped back onto the floor to grab their drinks, it became a chaotic scene resembling a broken conga line. People were running around the tables, sipping or spilling their drinks before jumping back onto the stage. Despite the chaos, they never stopped dancing or clapping.

It was very late when the band finally ended. Everyone was hugging and high-fiving. Their ears were ringing. They swayed back and forth as if the music still played.

Haruka told Liv and the others from the ryokan that she'd bring the van around the front. They all stood outside in the warm, humid air, talking about giant manta rays and the pig brain noodle soup. Cradling air sanshins across their bodies, they hummed loudly and poorly the tune from 'Shima Uta.'

Kasemchai hugged the girls before they got into the van and then waved goodbye. After the van drove off, his car pulled forward, and the driver jumped out to open his door. Kasemchai slid into the back seat, slumped against the armrest, and smiled.

He was conflicted. He had come to meet Liv and discovered Haruka, and now he was alone—but not lonely. He tried to name the feeling, and the word "satisfied" came to mind before he closed his eyes.

THE NEXT DAY, Kasemchai ran his fingers through his hair, dragged his palm over his face, and opened one eye. The other eye flickered before finally opening.

"Time to get up," he mumbled to himself.

He threw off the sheets, grabbed his AugWatch, and walked to a small table. A tray was laid out with a tall glass, a small carafe of iced coffee, a small container of sweet condensed milk, and a bucket of ice.

He flipped his hair with his left hand and added ice cubes to a glass with his right. Then, he poured condensed milk into the glass and topped it with coffee. After giving it a quick stir, he took the glass and sat down in a low chair on the veranda. He eased back into the chair, took a long sip, and sighed loudly before placing the glass on the table nearby.

His AugWatch sensed the lull, and Chanthavy's soft voice spoke up. "Good morning, Kasemchai. Would you like to see the news?"

"Yes, please."

Images appeared before him, and he swiped at them until he found one he wanted to watch.

Chanthavy interrupted him. "You have a few new messages, Kasemchai. One is from the woman you were with last night. Her name is Liv."

"Read it to me, please."

"It was great diving with you yesterday!" A dancing bikini animoji hovered in front of him, and then a fish animoji swam by. "Let me know if you want to talk about winning the Phoenix Global Water Challenge."

"What the hell is the Phoenix Global Water Challenge?" he muttered. With both hands, he vigorously rubbed his face, driving the sleep from his brain. Then he remembered the conversation about the electric tanker ships the previous evening. "Nope. Too much work. What I need now is a massage and a big breakfast."

He turned and faced the room off his bedroom. A massage bed had been placed in the room and draped with white towels.

A woman stood beside the bed. She bowed slowly and gracefully. "Good morning, sir. What kind of massage would you like today?"

Chapter Fifteen

It was 8 a.m., and the sun would be setting soon. Outside, it had begun to cool—one hundred-and-twenty-three degrees Fahrenheit.

Makoa stepped out into the hallway and stopped outside the door to adjust his hat. It was large, as all hats were these days, to protect him from the scorching sun. The hallway in the outdoor ring of his apartment was kept cool by the thick walls, and cool air rushed upward through the vents on the floor.

"Hey, Gene," he said to his AugWatch. "Play the latest *Look Up to God.*"

His companion responded, "You got it, man."

Philippe Labonne's voice enveloped Makoa. "God helps those who help themselves," he said. "So don't wait for divine intervention."

Makoa closed his eyes. The background music swelled almost imperceptibly. He felt a hand on his shoulder, jumped, and whipped around reflexively.

"You forgot your lunch, honey."

His wife, Anela, kissed him on the cheek and stepped back into the apartment. Makoa turned to continue down the narrow hall towards the elevator.

He inhaled quickly through his nose, and then his smile slowly faded from his face as he focused on Philippe's voice. His commute was so routine that he was only barely aware that he was outside and climbing the steps to board the elevated bus—a huge multi-car vehicle that straddled the streets on two sides. The passengers sat above the street while individual cars could pass underneath.

He searched for meaning and direction in the sermon. His mind wandered to the serenity prayer, which he often said to center his mind.

God grant me the serenity
to accept the things I cannot change;
courage to change the things I can;
and wisdom to know the difference.

What was Philippe saying? Was he telling him to change things or to accept them as they were? Could he accept that the last years were nothing but misery and horror? His brain started down a dark path. He thought of the ride to the airport on the day he left Hawaii forever.

The bus lurched. He looked around to get his bearings. Ten more minutes to his stop. He took another long breath, and this time, he smiled blissfully as he listened to Philippe's voice.

"The strength of water can shatter stone and bone. It carved great channels through the earth. Long ago, it took the form of enormous glaciers that crushed everything in their path."

The music swelled. Then, it faded to nothing.

"And yet, you can wave your hand through the water effortlessly. It trickles downstream, not moving the rocks, but moving around them."

The music swelled again. Then stopped.

"Imagine a cup of water. Place it in the hot sun, and what happens to it? It will disappear. Water is incredibly strong, and yet, when it encounters an obstacle, it doesn't stop. It moves forward. It always finds a path forward. It can drown a man as quickly as it can quench his thirst."

The bus lurched again, and the iron wheels below screeched against the rails. He stood in his seat and saw an old building was on fire.

"We're sorry, my friends," the bus's AI voice said. "There is a fire ahead, and we cannot proceed until the emergency is cleared. Please refer to your AugWatch to identify the next pick-up spot. Remember to stay in the shade."

Makoa was entranced for a moment by the heat waves made by the flames. The outside of the building blistered and expanded like a burning marshmallow, but there wasn't a single flame. It looked like a grotesque chimney with smoke and fire bellowing from the center.

At this time not a drop of water could be spared to put out fires, so buildings were often coated with an intumescent coating that swelled and created a protective barrier. The design prevented flames from leaping to other buildings. Most people believed it was a painless way to die as the occupants suffocated or passed out in the furnace.

The doors of the bus opened, and the stairways folded out. The heat of the outside invaded the bus cabin instantly and people hurried to get out before it turned into an oven.

"Gene, how long is the walk to work?"

The voice from Makoa's AugWatch responded, "You know what, man? It's only eight minutes from here. Just follow the arrows."

Makoa started down the sidewalk. He hated it when he entered the winery sweating. The clamminess hung on him all day.

Chapter Sixteen

KASEMCHAI SAT in front of his experience sphere a few days after returning from Ishigaki Jima.

"Chanthavy," he said. "Show me the latest news."

"Of course. The first story I've selected is from Canada. You may find it interesting because of your family's business in the region."

A news anchor appeared. He was sitting behind a desk, and an image hovered behind him of a truck pulled off to the side of the road. A man lay down in the dirt beside the truck, one arm outstretched, the other under his body in an unnatural position. He was not moving. The image from the background zoomed into the foreground.

Kasemchai reached out and dragged the corner to make it larger. He stared at the man on the ground.

"For unknown reasons, January water shipments were down by three ships, and February water shipments were down by five ships. Gang activity has increased dramatically, and hijacking municipal water trucks has increased by nearly thirty percent. The government is increasing truck security but has struggled to keep up with the violence erupting around the country."

Kasemchai leaned forward. "Hey, Chanthavy, stop the news, please."

The news disappeared, and he stared across his living room.

"Thank you, Chanthavy," he said as he sank back into the couch.

"You're welcome, Kasemchai," his companion said. "The news is very intense. Would you like to take a moment and focus on your breathing?"

"No thanks, Chanthavy. Is Angélique available?"

Kasemchai heard the repeating tone when someone was being called, and then Angélique appeared before him. Her avatar looked around and sat down in the chair across from him. He knew his AugWatch had mapped his room and made it appear the avatar was sitting on the chair, but it still made him marvel to think the computer was putting on this show for him. *They think of everything*, he thought to himself.

"Hey, Kasemchai! Hold on one minute." Angélique turned and disappeared as she darted off to the left. She reappeared a moment later. He could tell she was leaning over her AugWatch disc, allowing it to scan the necklace so she could add it to her avatar.

"I love them! What do you think?" She turned back and forth, revealing the pearl necklace around her neck.

"They look great on you. I found them in a tiny shop near Kabira Bay."

"The black pearls are so unusual. You're very naughty. I was joking when I asked you to pick them up for me."

Kasemchai laughed. "Hey, I do what you ask of me."

"So, how was the diving?"

"Incredible!" Kasemchai leaned forward to share some of the photos. He thought of Liv and Haruka and then decided against it. "You were right. It gave me a chance to disconnect. The ocean is crystal clear. I didn't think I'd like the food, but after a long day of diving, it really hit the spot."

"You're leaving out the most important news," Angélique reminded him.

Kasemchai's face flushed with heat. "I am?"

"Did you see any rays?"

"Oh, oh, sorry. Yes, the giant mantas were incredible. They were so enormous!" He leaned back and spread his hands out as wide as he could. “Right after I dove into the water, I almost landed on top of one. It was so big I thought it was the ocean floor at first.”

“That’s incredible!”

“And we went down to these coral mounds and hung on, waiting for them to fly over us, just like fighter jets.”

“Speaking of fighting,” Angélique said, switching topics. “Are you watching the news? People are getting all worked up again about the price of water.”

“I know. And your father was on TV last night telling everyone to save their water and use only three liters daily.”

“Almost, Kasemchai,” Angélique chided him, “He said, ‘Treasure the holy trinity: God, the Father, and holy water.’ He didn’t literally mean to drink three liters a day—that’s the newscasters mangling his words. He’s just saying that God would want us to treasure every drop.”

“I know.” Kasemchai laughed. “And as usual, my father said the opposite.” He held his fist up and gesticulated with his finger extended. “Water is your right! Drink the water! Splash around in the water! There will always be more.” He sighed. “Can’t they ever get on the same page?”

“Of course not. It’s our business to disagree,” they both said in unison.

“It’s good business to make bad politics!” Kasemchai added.

They both fell back, laughing.

Kasemchai slouched on the couch and caught his breath. He wiped a tear from his eye and said, “Well, at least they’re doing their best to save the world.”

“How’s that?”

Kasemchai paused and pulled his hand away from his face but didn’t drop it. “I mean, they may disagree on politics, but they both want what’s best for the world. They’re just trying to bring water to everyone.”

Angélique roared with laughter. “Is that what you think, Kasemchai? That they want to save the world? You got the *world* part right, but they want to rule the world.”

“You’re being a bit cynical, don’t you think?”

“Do you think our fathers care about the price of water? They only care about profit. They run a business everyone needs, and they have a complete lock on the source. They're just trying to get a bigger piece of the pie."

"What do you mean?" Kasemchai could feel his face warming. "They're doing everything they can to ship more water. They don't need a bigger market. They don't need more customers. They just need to get water to people faster."

"No, Kasemchai. You need to open your eyes. We don't know why the water falls in only two places—"

"Uh, duh, it's all the pollution."

"No. What I mean is that no one predicted that would lead to all the rain falling in two places. It wasn't gradual. It just happened. These things are unpredictable, and they could change again. That's what our fathers fear more than anything else in the world."

Kasemchai got up and began to pace. "Look, our fathers have more money than they know what to do with. The rain could go back to normal, and they would be just fine. They could just sit back and drink their wine."

"You don't see it, do you? That may have been your father a long time ago, but now they're different. They don't want things to go back to normal." Her voice was soft but cut through the silence like a knife. "They *like* this world they've carved up between the two of them." She leaned toward Kasemchai. "And besides, what about you and I? Don't you think about the empire they're going to give to you?"

"I don't understand what you're saying. Who cares about an empire? I have everything I need: my cars, my investments, and the AquaSpire. You're trying to make our fathers into something they're not."

He was talking very loudly now. Angélique just smiled at him, like the Cheshire Cat from Alice in Wonderland.

"What are you smiling at?" demanded Kasemchai.

Like a cat with a bird firmly under its paw, Angélique was in no hurry to respond. She looked off to the side. "Look, maybe that's just me being cynical. It's in my nature."

Kasemchai's eyes drooped, and he rubbed them vigorously with the palms of his hands. He was trying to process what she had said, but it was so unexpected. So caustic. "Yeah. Anyway, I have to get some stuff done. I'll catch you later."

"Thanks for the necklace, Kasemchai. It's precious." Angélique waved goodbye and then disappeared.

THE NEXT WEEKEND, Kasemchai drove to his parents' estate for lunch. As he walked through the house, he tried to talk to the staff, but he was still mulling over his conversation with Angélique.

He had never considered his father anything but good, and he thought the same of Philippe. Angélique had introduced a dark stain that slowly crept into all of his memories.

"How's it going, Kasemchai?"

Kasemchai realized he was standing in front of his father. He smiled —poorly. "I'm fine." It sounded like an excuse more than a statement.

His father stared at him. His eyes flicked back and forth as if he were trying to decipher what was going on in his son's head. He placed his hand on Kasemchai's shoulder. "Come on in, Kasemchai. Let's have some lunch."

"Sounds good, Father. I'm famished." It was a lie.

AFTER LUNCH, Kasemchai strolled around the grounds and stopped in front of the waterfall in the south corner. For a long time, he simply stared at the water; his vision blurred, and he only registered the white noise of the water falling on the large river stones below.

When he walked back inside, he passed his father's office. He couldn't see who was projecting up from his father's AugWatch disc, but he recognized the voice immediately: Nary, the head of transportation.

"OPEC will make their statement on Friday. They will cut production by two million barrels a day next month." Nary had been beside Preap since he had taken over his first ship, and he could talk to Preap like no one else could. Their arguments, loud and raw, weren't born of hostility. The two men trusted each other implicitly, and their passion for their convictions was as powerful as their ability to reach a consensus.

Preap turned his head and stared sideways at the wall. His jugular vein stood out prominently as he clenched his chin. He snapped his head back. "They're cutting production now? Bunker fuel will soar! Higher water costs create fear. When people fight, they damage equipment." He turned away and blew out between his pursed lips. Then he turned back and pounded on the desk. "OPEC isn't an organization. Why don't we call them what they are? The *Cartel* of Petroleum Exporting . . ." He could only get the letter "C" out of the last word, "Countries," before his face contorted into a snarl.

"I know, I know, Preap. I think they want to remind us of their position. Discourage us from investing too heavily in oil exploration."

"Schedule a meeting. I want to meet with the Saudi king directly."

Kasemchai hesitated, then knocked on the door. His father waved him in. When he walked to his father's side, he could see Nary's avatar sitting on the chair across from his father.

"Hey, Nary. How are you doing?"

Nary had a surprisingly full head of hair for a man of his age, and he broke into a wide grin when he saw Kasemchai. "I'm doing well, Kasemchai. Good to see you."

Kasemchai leaned on the credenza and stretched out his legs before him. His father swung his chair around to see them both.

"I have a question for you guys," he said, pausing to gauge their interest. "I know I don't know your business as you do, but what if you could significantly reduce your use of oil for shipping? What could you do with that extra profit?"

"Well . . ." His father looked up at the ceiling for a moment. "If there was a reliable source of energy that we could use in our tanker ships that cost less than bunker fuel—mind you, bunker fuel is the cheapest fuel we can buy—then we could lower the cost of water."

"You wouldn't just keep the profit?" asked Kasemchai.

"We'd have to use some of those gains," Preap said. "You know how a business operates, Kasemchai. But eventually, we could lower the price of the water."

"And we'd reduce our reliance on oil suppliers," Nary pointed out.

Preap laughed. "Then we'd have a new master, Nary. Would they be as benevolent as our oil masters?"

The two laughed heartily. Kasemchai pretended to laugh along with them. He thought about his conversation with Liv on Ishigaki Jima. "I ask because I've been exploring investments in a technology that can run tanker ships with solar power. This would dramatically reduce your reliance on oil to move water."

His father glanced up at Kasemchai. "Solar? I'm sorry. I don't like being in this bind with the Saudis, but fuel is cheap. And it works day and night. Besides, this is not a new idea. Lots of companies have tried to make solar panels for ships. They're just not reliable."

"That's just it, Father. There have been major breakthroughs in solar panels. They're way more efficient now. What if we could produce thirty percent of the energy you need on a ship? You'd use less fuel, and the ships could run faster."

Preap inhaled through his nostrils, then let it out in a long, loud exhale. "What we need is to remind the Saudis who have the power. They have oil, but oil is nothing without water to drink. They will listen to me."

"I agree with your father, Kasemchai. I know that you're a savvy investor. My son uses your dating PrefX often," he chuckled. "Maybe too often. But the Saudis only understand pressure. It's how they made gold from sand."

Kasemchai stared at the ground and gathered his thoughts. Preparing for his real question. "Father, I'd like to join your meeting with the Saudi king. You've seen the success of my negotiations with investors for my PrefXs. We've talked many times about how I could become more involved with your business—"

"You know there's nothing I'd love more than to have you become more involved," Preap interrupted him. "There are ways, and your tech background would help us modernize our business. But this is not the place to start. The king is a sophisticated negotiator."

"Yes, but what if my solar idea worked? You could save money on shipping while shipping more often. That would be both a stick and a carrot for the Saudis. And the Labonne family, if you increased your shipments to the US by twenty percent, that would give you an incredible advantage. You could take market share away—"

Preap slammed his hand on the desk. "Look, I don't care about increasing market share. The only thing I care about is bringing as much water as possible to our side of the world. Shipping water is a human necessity, not a consumer business, Kasemchai."

Kasemchai's ears burned. He was mixing ideas. The solar, growing the business, joining the negotiations with his father. He threw out everything and won nothing. "Sorry, I didn't mean to imply that."

Preap sighed. "That's okay. It's a tricky business. Tell me, didn't you just return from a diving trip?"

Kasemchai knew his father was changing the topic. The discussion was over. The sooner he ended this conversation, the sooner he wouldn't have to look like a child in front of Nary. "I did. Diving with the manta rays in Ishigaki Jima."

"Ah, incredible. I envy your life. Did you go with anyone?"

Kasemchai smiled. "Angélique was going to meet me there. It was her idea. But in the end, she couldn't make it."

Preap glanced at Nary and then back at Kasemchai. "Ahh, Angélique. Such a nice girl. It's so good that you guys stay in touch."

Kasemchai excused himself from the room. He looked up when he saw Kosal waiting for him outside and then dropped his gaze. "Bring my car around, Kosal."

"Of course, Mr. Seng. I have already done so."

They walked together in silence. As they neared the main door, Kosal scooted ahead to open it. "I heard rumors that you're investing in another company. I can't imagine what it's like to be in a boardroom with you, sir. I can only imagine how intimidating you must be."

Kasemchai stopped short. He stared down at Kosal. "Were you listening to my conversation?"

"Sorry, sir? I don't know what you mean. I only read about these things in the news."

"That's not what I mean." Kasemchai stared at him. His eyes flicked left and right as thoughts darted through his head. Finally, he let out a deep sigh through his nostrils and started towards the front of the house again. They walked down the stairs towards the car.

Kasemchai stopped and looked at Kosal. A wane smile could not hide his frustration as he stepped into the vintage Aston Martin Valhalla. He fired up the car and gave the engine a few quick revs.

Kosal walked back to the stairs and turned to bow. He would stay bowed until Kasemchai was gone.

Kasemchai muttered out loud, "Generations will remember me for my dating PrefXs."

As he drove, he tightened his grip on the wheel and navigated the circular driveway. He remained composed as the road opened up and curved towards the lush jungle ahead. When he felt he had put enough distance behind him, he pressed down on the gas pedal. As the car hurtled forward, he was pleased to hear a satisfying *pop* from the exhaust as the engine screamed and revved up to over six thousand RPM.

He muttered under his breath, not unlike Liv. "Kasemchai, playboy son of Preap, the king of dating PrefXs, negotiator lightweight. Is that how people will remember me?" He gripped the steering wheel harder. "The answer is simple: I need to create a real business. Something tangible he can see."

Chapter Seventeen

WHEN WATER WAS FIRST SHIPPED from Asia and Africa to other continents, it was difficult to distribute it beyond the ports. Infrastructure further inland deteriorated, and people who could move left for the cities near the ports. High-density housing sprang up to accommodate the vast number of people who swarmed the ports. It had to be easy to construct and cost little to maintain.

They referred to them as termite towers. Makoa's family lived in buildings designed to replicate the natural cooling systems found in termite mounds. The walls of these structures were thick and constructed from mud-colored concrete and brick. The exterior was rough, featuring small windows to minimize heat absorption. Inside, the center of the building was spacious, with small patios for each unit facing inward and shaded from the sun.

Square-shaped cooling towers with tall slits stood on the top of the building's roof, where the air was cooler and less dusty. The building itself could breathe, with the windward side pulling warm air down, cooling it as it circulated below ground. The cool air then ran up through the building, pushing the warm air out through the leeward side of the chimneys. Even the roofs were uniquely designed to keep the building cool, slanting inward to capture any moisture and directing it down long rain chains toward the courtyard. Everyone still spoke about the ten minutes of rain they received four years ago.

They were ugly, with dark interiors and exteriors caked with dust and bits of trash stuck to the crevices. However, no one could deny that these housing units served their purpose as cheap and cool living spaces.

Makoa's family had been preparing for the dinner party all night at their home. It was 2 a.m., and the guests would be arriving soon. The refrigerator was full of different vegetables, white rings of squid, and marinated cultured meat strips. Anela had already fried up the crispy lumpia rolls filled with carrots, green onions, and lots of garlic.

She was now working on her soil stew. The name was a dark joke. In harder times, root vegetables like carrots and parsnips were all they had, and they couldn't waste water washing them. It was mixed with roasted crickets and mealworms for protein. The earthy flavors of the vegetables complemented the nuttiness of the insects. If available, they added herbs for depth.

She dipped a spoon into the brown stew and tasted it. *Just a few minutes more*, she thought to herself. She spoke to her digital companion, "Young-Soon, play a five-minute song. And add a gong to the end."

"Of course, Anela. Something with a ukulele?"

"You know me so well," Anela said.

The song started slowly with the soft strum of an ukulele. A woman's voice joined in and sang about waves and seashells.

"Are you making soil stew? I smell rosemary and thyme," Young-Soon said.

Anela looked down at her AugWatch. She scrunched her mouth and nodded. "You have a sensitive nose, Young-Soon."

Apikalia shrieked as her older sister chased her through the kitchen.

"Girls," scolded Anela. "Be careful."

Kailani looked up at her mother and then sideways at the electric pot. "Soil stew?"

Anela glanced down at her daughter for only a second before scanning the patio again. "It is, honey."

"I'm hungry."

Anela smiled at her two girls. She knew they wouldn't relent until she gave them a bite. It made her smile to watch her little Apikalia, too young to fully understand what her older sister was asking, but determined to get her fair share.

"One bite." She waved her long spoon at them, pretending to look stern. The two girls exchanged glances and giggled, confident they had tricked her. "You each get one cricket."

She lifted the lid and spooned out a single cricket and a slice of carrot, blew on it, and lowered the spoon to her daughter.

She greedily ate it, and Apikalia began to chat, "Me! Me!"

Anela repeated the process for her younger daughter, then scooted them out of the narrow kitchen. "Come on, you two. Go play."

She looked at Makoa to see he was snapping a tablecloth over the fold-up table, which extended the dining table. Folding chairs in various styles and colors circled the tables.

"When are you grilling? Everyone should be here soon," she called across the room.

"Just a minute. I want to bring out the cups."

"Hah, I know you just want everyone to see you so they know you can cook."

Makoa pretended not to hear her. He pulled out a box from the closet filled with slender glasses and stainless steel cups. Each glass was placed into a stainless steel cup that was wider but half as tall. After arranging them on the table, he walked up to his wife and hung his arm around her shoulder. "You're just afraid they won't know who the real cook is around here."

He gave her a squeeze, then just as he started to walk away, he looked down her back and gave her bottom a smack.

Anela turned and swiped at him with the towel she had in her hand. Her face broke into a broad grin. "You watch yourself. That's how we ended up with a second daughter. You looking for a third?"

Makoa looked back toward the bedroom. "We have time."

"Get away from me," Anela huffed

Makoa turned and started the electric grill, a grin lingering on his face. Soon, the large squid began to sizzle and teriyaki sauce slowly spread to the sliced onions and turnips, building a delicious smoky and sweet aroma. He sprinkled a dash of salt and carefully placed the grilled squid onto a cutting board. He was nearly finished cutting it up when he heard the doorbell ring. He wiped his hands on a towel before Anela went to open the door.

Family and friends spilled into the house. As each group arrived, people stood to embrace the newcomers with a hug or a friendly slap on the back.

Anela and Makoa greeted them with, "Are you thirsty?"

The guests responded, "I am not thirsty."

Anela gave each guest the nested cups, and Makoa poured cool water from a carafe. The water overflowed from the tall glass and into the second cup, a tradition that emerged during the drought. Filling a glass up to the top was a sign of generosity, but no one would be so foolish as to waste a drop of water. People began to place a second cup under the first and enjoyed a little extra water that way.

The little apartment quickly became crowded, and people moved around, trying to find a place to sit. When Makoa's older sister U'ilani arrived, she loudly announced that she had broken up with her latest boyfriend and inserted herself between Anela's cousin and his wife at the dining table. When seats ran out, they stood. Water turned to hard kombucha, and the room was filled with stories and laughter. They spoke about how many tanker ships of water had arrived and how many were coming. The girls ran around between tall legs, popping under the dining table and reappearing on the other side.

A tall man with huge arms carefully squeezed through the front door. "Uncle Keone!" shouted the girls. "Play a song! Play a song!"

Keone bent down to pick up the girls, one in each arm. He hugged them and kissed their heads, making them giggle with delight. "Be patient, girls. Let me eat first."

Kailani ran to the table and grabbed two lumpia. She pressed them into his hand. "Here, eat them. Are you full now?"

Keone laughed. "Not yet, baby. You'll know I'm full when you hear my ukulele."

At dinner, the tables were mismatched heights. The tablecloth was hidden underneath a mad collection of platters, glasses, and plates. There were bowls of bright green wakame salad, crisp and tangy. The hijiki salad was savory, with stringy hijiki seaweed tangled around plump edamame beans. Cool pickled jellyfish was served in glass bowls. Warm teriyaki squid was carefully layered on a long platter and served beside bowls of grilled root vegetables.

Makoa asked everyone to clasp hands, and the room became silent. He led the group in prayer, starting by saying, "Ola i ka wai," which meant "water is life." He thanked God for bringing the family together on such a joyous day. For giving them work, not as much as they needed, but more than they had had in the past. For the family that they had in Hawaii, and even for President Bob. They knew he didn't believe in God, but they would pray for him anyway. That drew many amens. They lowered their heads in one final moment of self-reflection.

The serenity transitioned to noise and joy as plates filled with food. It was not easy to eat as they often bumped elbows with their neighbors, but that only led to laughing apologies and bad jokes.

Anela stood and walked to the patio off the living room. She opened the door, reached down, and picked up two baskets. She turned back to the room and called for everyone to be quiet. "Girls, I just saw something on our patio. I think it was the Easter Bunny."

Kailani shrieked and clapped her hands. She immediately grabbed the basket from her mother. Apikalia held her hand in front of her face. Her eyes were wide.

"Don't worry, honey," Makoa said. "Let's go together."

The girls whipped around the patio, grabbing candies to shove into their baskets. When they were certain they had found every last sweet, they returned to the living room floor, and the older cousins helped them sort their loot.

Keone moved a folding chair into the living room and sat on it with his ukulele. For a long time, he just played. The room was quiet, and people swayed, bathed in the magic glow of ancient Hawaiian music.

After several songs, he began to sing a classic, "White Sandy Beach" by Israel Kamakawiwe'ole.

I saw you in my dream
We were walking hand in hand
On a white sandy beach of Hawaii

U'ilani stood up. She swept her thick black hair behind her shoulders and began to dance. She swayed to the left, then back to the right. Her arms were outstretched, and her hands moved fluidly—a single wave from her shoulder to her fingertips. She stepped out with her toes and back again, slowly rotating.

Anela watched her daughters stare with open mouths. She reached out her hands and pulled them up. The three joined U'ilani on the floor to dance to Keone's ukulele.

On the tenth floor of a dirty brown building on a withered and dry landscape, a little piece of Hawaii was passed from one generation to the next.

Chapter Eighteen

Liv's office had tall windows, which must have let in lots of natural light when the building was first made. But no one worked during the day anymore, and the night sky was too hazy for anyone to see the stars. The windows were now more like giant mirrors.

The room was spacious, with dark blue flooring and clusters of three to five chairs surrounding a low, round desk. The chairs tilted back, and some had large, swooping wings that isolated the head—a soft interpretation of the Fritz Hansen Butterfly Chair.

In the center of the room sat a small gathering area designed like an A-shaped cabin featuring comfortable green benches. Lacquered tree trunks rose up from the floor and vanished in the ceiling, creating a cozy forest atmosphere.

People worked in groups around a small table where they rested their AugWatch discs. Some had pulled their chairs back and were standing, looking left and right, moving forward and back in a virtual or augmented world of monitors and data models.

Others lounged in chairs, legs stretched out or tucked under. Their eyes darted quickly, and at other times, they stared at a spot in space. Or they had animated conversations with someone only they could see.

They deftly flicked their fingers or pushed and waved their hands in front of them. Each chair looked clean, except one.

The floor around Liv's chair was cluttered with plastic ship models, black squares of paper taped onto parts of the deck. Two different-sized boxes lay beside her. A shiny tablet with an electronic pencil was on top of one of them. Three tea mugs, all empty, sat on the center table next to a large empty bottle of awamori.

Johannes leaned over her chair from the side. "Hey, Liv. How's it going?"

Liv looked up at him and waved her hand through the air so that she could see him better through her digital workspace. "Fine. Just reviewing numbers for the battery test."

"Good. Good. Do you think it will be finished today?" asked Johannes.

"No problem. I was just reviewing for inconsistencies. But I'm getting close. They'll be ready today."

Sophie sat up in the chair to Liv's right. "Are we going to lunch soon?"

"Sure, I can go in ten minutes. Are you joining us, Liv?" asked Johannes.

Liv sank in her chair, "I don't think I can. I need to work on this report if you want it today."

Her AugWatch vibrated, and she glanced at her digital workspace. She tilted her head slightly and furrowed her brow when she recognized the caller. With her index finger, she swiped the air to take the call while Johannes and Sophie continued discussing restaurants.

"And you want me to come upstairs?" Liv said. A pause, and then, "Okay, I'll come up now."

"What was that?" asked Sophie.

"It's Luuk. He said there's a new investor, and they want me to come up and talk to them."

"Why would they need to talk to you?" Johannes said, sounding a little accusatory.

Sophie gave Johannes a gentle push.

"I don't know. He just said come up now," Liv said.

"Ignore Johannes." Sophie gave Liv a reassuring tap on her knee. "They probably have a good question."

"Yeah, like how to change the battery on the experience sphere." Johannes laughed at his joke.

"Better go, Liv." Sophie scowled at Johannes and stood. She held her elbows out to the side and balled her fists in front of her, then twisted to the left and to the right, stretching her back. "I'll get something for you. What do you want?"

"Thanks. I'll have fish sausage." She turned and stared up at the stairs and said with a distant voice, "Curry if they have it."

After they left, Liv got up and headed towards the staircase. The conference room was located one floor up in the far corner. Through the glass walls, she could see Luuk speaking with three men who were facing away from her.

She leaned in through the doorway, keeping one foot firmly planted in the hallway. "Hi Luuk, did you want me?"

"Thank you, Liv. Please come in." Luuk looked more excited than usual. "I'd like to introduce you to our new partners. Liv, this is Mr.—"

"Hello, Liv. It's very good to see you again," Kasemchai said. The twinkle in his eye betrayed the subtle smile on his face.

Liv stared in amazement. It had been a month since she visited Ishigaki Jima. At first, she felt hopeful that Kasemchai might respond to one of her texts. Then foolish for thinking she could text someone so powerful.

Luuk's mouth hung open, and his eyes narrowed. "You know each other?"

"Of course, Mr. Visser," Kasemchai said. "I'm afraid I owe you an explanation. It's true that I have been researching your company, and it's true that I see a huge market for solar-powered tanker ships. But the real reason I'm here is Liv."

He walked towards the window and turned back to face the room. He smiled, relishing the uncomfortable silence. "It was chance that we crossed paths. And Liv did not waste the opportunity to simply say hello. She pitched me on why this company could reduce the energy costs of tanker ships. And my family wants to save money on bunker fuel."

"I see." Luuk didn't look convinced that Liv could have had any such conversation with Kasemchai, but words tumbled from his mouth. "Yes, she's an invaluable part of our team. I had no idea, Liv." He paced back and forth, then he stopped. He smiled at Liv. "I had a good feeling when we hired you, Liv. But I have to confess, not in this way."

Then, as if he suddenly remembered where he was, he stopped and looked at Kasemchai. "So, how did you want to proceed?" he asked.

Kasemchai regarded Luuk. "I know that you're the genius—the founder. I want you to have the freedom to hire the people you need. To drive the projects, you need to make this a reality. I don't want to stand in the way of your development." He walked over to Liv and then turned back to face Luuk. "I'm investing my faith in you, but I need to have someone who I can check in with constantly to update me on the company—someone who understands your work intimately and can help me understand your progress. You will stay as CEO, the visionary. I would like Liv to be made your president."

Luuk's mouth opened, then closed again. Then, it opened again. "I can't just put Liv in charge of day-to-day operations! She's an engineer—not even a senior engineer."

"Mr. Visser, I believe in your vision. It has great potential. But I believe in Liv to get things done. To take you to the next level. Can you make this happen?"

Luuk gaped at Liv, clearly trying to find the right words. He eventually turned to Kasemchai. "Yes, let me figure this out. I'll have an answer for you in a few days."

"Great. My attorneys will share the details of our investment with you." Kasemchai turned. "I'm famished. Do you think you could join me for lunch, Liv?"

Chapter Nineteen

KASEMCHAI'S CAR stopped in front of a home midway down the block in the Kronlingen-crooswijk district. The driver let him out. One of his security detail from the car that followed them stepped over to Liv's side and let her out.

Liv walked with Kasemchai towards the door. "What are we doing here? I thought we were going to lunch?"

"We are."

She glanced to the side. The streetlights disappeared into darkness.

"Just wait." Kasemchai stepped up to the door and knocked, then stepped back.

The door was opened by a man wearing a very fitted dark grey suit, with a bright rouge tie and shoes polished to a mirror finish. He smiled broadly upon seeing Kasemchai. Then, he bowed slightly with his arms out.

A sinewy woman stepped forward from behind him. She wore a white short-sleeved shirt with the sleeves rolled up, showcasing her heavily tattooed arms. A clean, starched apron adorned her waist, and her hair appeared electrified, held down only by the top of her skull. She threw her hands up in the air. It looked like they were singing the "YMCA" song.

"Mr. Seng, welcome!" they shouted in unison.

Liv and Kasemchai followed them into the house. Inside, it was surprisingly large. The ceiling was domed and painted like a beautiful blue sky with lazy clouds. The walls were smooth grey cement. Ivy hung down in many colors and forms. Thin steel rods emerged from the floor, close together but swaying slightly in the breeze, obscuring each of the tables and their diners.

"A secret restaurant?" she asked.

"Not that good a secret. I found it," Kasemchai said.

The maître d' and the chef walked them toward their table. Kasemchai shook hands with the chef.

"Are you thirsty?" asked the chef.

"No, I'm not thirsty," Liv and Kasemchai responded in unison.

"Very good then. I am the executive chef, Arlo," she announced to Liv.

"Nice to meet you. I'm Liv."

"Excellent. When I learned you would be joining us tonight, I thought deeply about a meal that would surprise and comfort you. But let's not get ahead of ourselves." She raised her palm upward towards the maître d'. "Please, welcome our guests. I will return shortly."

"Of course, Arlo."

The maître d' pulled the heavy chair out for Liv, and she sat. Then he hurried to Kasemchai's side to pull his chair out for him. "Welcome to Watercress. Liv, may I take a brief moment to introduce you to our restaurant?"

Liv nodded in the affirmative, tilting her head to the right. "May I ask a question?"

"Please, of course."

"What is a watercress?"

"Aha, I'm not surprised at all that you have not heard of this plant. It's native to Europe and Asia and must be grown in the wettest environment. It's said to be the healthiest vegetable in the world, with a slightly piquant taste." A cart of lacquered wood with brass fittings was wheeled up behind him by a server. The maître d' placed his hand on the cart. "And that tasty little plant, not seen around here for years, is the inspiration for our menu—water. Our menu is a song of our love for water. Simple, exotic"—he leaned in just a bit to smile—"and perhaps forbidden. You will experience it all today."

He turned, opened the cart, and pulled out a slender carafe. "Before we begin, I must beg your forgiveness, Mr. Seng. I'm certain this is nothing like the water you are used to drinking."

Kasemchai gave a warm smile. "Not to worry. I always appreciate a good glass of water."

"Very good. This water was mined from the ice shelf below Antarctica. It's thousands of years old, perhaps more. It was brought to us in its frozen state. Slowly melted under sunlight, then triple charcoal filtered."

Liv visibly sat back and blinked rapidly. "Uh. I thought there was a moratorium on removing ice from Antarctica."

"There is. And just this morning, it was defrosted." He pointed towards the door in the back of the room with his outstretched arm.

Two servers hurried out to their table. They placed a glass in front of each diner.

"The ice crystals are created by rapidly turning the bottle in an ice bath," the maître d' said.

Kasemchai lifted his glass. He waved it under his nose. Then, took a deep sip. "Delicious! This may be better than the water from home."

The maître d' threw up his hands in the air. "Please, sir, I appreciate your humoring me, but I have been to Cambodia to taste water directly from the wall. There is no comparison."

Kasemchai laughed. "Don't worry, I am known for my generous tips. I don't need to be embarrassed in front of my guest."

"Thank you, sir!" He snapped to attention and then, with a conspiratorial voice, said, "I will focus on the reason you have come here. Let our journey begin."

He bowed and strode towards the kitchen.

Liv looked over at Kasemchai in wonderment. "How did you know of this place, Kasemchai?"

Kasemchai removed his linen napkin from the table, unfolded it, and laid it over his thigh. He grinned while smoothing it out with one hand absentmindedly. "When you are very rich, where do you want to eat, Liv?"

"I don't know. Probably the newest, hottest restaurant," answered Liv.

"That's right. The hottest restaurant. The trendiest restaurant. Somewhere where everyone will see you have fabulous taste." He waved his hand around. "No windows. So, where do you want to eat when you're very, very rich? Crazy rich?"

"I don't know, somewhere exclusive?"

"The answer is, you just want to eat. You want to eat something special. The only people who should see you eating are the server and the chef, so the chef can tell you why the food you're eating is so special. An intimate moment. A direct connection between you and the kitchen." He lifted his glass of melted Antarctic ice and waved it toward the direction of the kitchen. "The magic that happens back there."

Liv laughed. "So we're in for a magical experience?"

"We are, Liv. And it started with meeting you."

Her cheeks flushed and Liv averted her eyes.

"Tell me, Liv." Kasemchai leaned forward and rested his elbows on the table. "I have done my research. Sol-vang has the best technology for solar-powered tanker ships. But what am I investing in?"

Liv's face dropped and her back went rigid.

"No, no, no." Kasemchai raised his hands in the air. "This is not a test, Liv. I want to hear from you. Why do you personally believe in Sol-vang? Why do you think it's so special?"

"Ah, of course." Liv sat back. She smiled, tilted her head to the right, and stared up toward the left, deep in thought. "Well, most tanker ships today run electric engines to turn the propeller. They generally have two or more diesel-electric systems that drive an electric generator. One generator is used to power onboard systems, like lighting. The other is used when the ship is underway to power the electric propulsion motors driving the propellers."

She placed the disc from her AugWatch on the table and created a model of a tanker ship that hovered over it. It was slightly translucent, and the engines and generators appeared in solid color inside the ship. She then added solid color panels on top of the ship.

"We are developing a solution that supports the engines." She swiped her hand, and the smaller engine disappeared and was replaced by a tall rectangle. She pointed to the two solid panels on the top of the ship. "We can remove the smaller generator that creates power for onboard systems. Solar panels would charge the batteries." She pointed to the tall rectangular shape next to the engines. "Underway, the energy stored in the battery that was not needed for onboard systems would contribute to the power for the propellers, reducing the total amount of bunker fuel needed to run the ship."

"Fantastic," Kasemchai playfully slapped the table. "So we just need to find customers who want to pay less for shipping."

"Well, it's not that simple. These are still big batteries. And we're replacing a small electric engine with solar arrays, a charge controller, an inverter, and a battery. It's still complicated and expensive."

"So, why does this matter to you, Liv?" Kasemchai asked.

Her eyes twinkled with intensity, "Because I am very, very passionate about batteries."

Kasemchai laughed. "Really, batteries."

Liv rolled her eyes over the ceiling, scanning her brain for something that Kasemchai could relate to. A wicked grin came across her face when she found it. "Let me compare this to something you're familiar with. You love fancy cars, right?"

"Sure."

"We've known for a long time that while cars are useful and fun, they also create combustible gasses, including carbon dioxide, that are extremely unhealthy to humans. It's also why the planet is so hot right now."

Kasemchai leaned on one of his chair's armrests. He stretched his neck out. "Well, it's also hot because it pretty much only rains in two parts of the world."

"Of course. But why did our environment change so dramatically? Burning oil produces carbon dioxide that hangs in the air like a lid on a frying pan. The Earth creates warming infrared waves that get trapped by the lid and bounce back, making it hotter and hotter."

"And that's the reason why it rains in only two places," stated Kasemchai, punctuating his statement with his open hand.

"Why is that?"

"Because the condensation runs down the lid and falls down the edges."

Liz giggled. "Well, not exactly. But I suppose, in a way, that's exactly what's happening. We knew we were sitting in a warming frying pan, so people started to use batteries instead of oil for things like cars, but we still made a lot of lithium-ion batteries. And these batteries can explode in extreme temperatures." She stopped and leaned forward. She lowered her voice like she was delivering a ghost story. "In the twenties, a ship sank with four thousand luxury vehicles, Porsches and Lamborghinis."

Kasemchai's eyes shot open. "That's tragic!"

"Yes. But why did it sink? They believe a fire started in one of the car's lithium-ion batteries. It overheated and overheating can lead to thermal runaway, gasses build up and . . ." She held her hands apart and, louder than she had intended, said, "Boom!"

Kasemchai startled then laughed at his surprised reaction.

For a brief second, she marveled at the thought that she was sitting across the table from Kasemchai Seng, and truth be told, he was far more handsome in person. She also knew she was getting animated, but she couldn't slow herself down. When she talked about batteries, something took over her. "But even that wasn't enough to change how batteries were made. Companies are experimenting with safer lithium-sulfur batteries or iron flow batteries that are literally stored in shipping containers. But we're working on something that's more energy-dense, a sodium-ion rechargeable battery."

On the table was an ornate silver salt and pepper server with oval-shaped glass bowls and tiny silver spoons. She scooped up the salt and let it gracefully fall back into the bowl. "Salt. It's everywhere. It's made from seawater. And sodium is inert, meaning it won't overheat or explode."

Liv paused and breathed in deeply through her nostrils. She reached out to her water glass but realized she was too energized and might spill it. Instead, she ran her finger around the cool bowl, resisting the urge to lick the condensation running over her fingers.

Kasemchai didn't seem to notice her nerves and instead took the opportunity to jump into the conversation. "Okay, so you have a solution that removes a small generator and replaces it with solar panels and a battery made from salt that won't blow up ships. What am I missing?"

"It's the combination of the three. A new solar array, removing the engine, and a sodium-ion battery that can fit in the remaining space costs the nose of a salmon."

"Excuse me?"

"What?"

"Why are we talking about salmon? You said it costs the nose of a salmon?"

"I'm sorry. 'Het kost het neusje van de zalm.' It's a Dutch expression. It means something is very expensive."

"Ahh, always the water. Why are you focused on doing this for tanker ships? Why not airplanes? Surely that's a bigger market?"

"Sure, Kasemchai, and they create more greenhouse gas emissions than ships. But most of the world's transportation is carried by ships. Ships are the only way to deliver water to much of the world. Just fifteen of the biggest ships emit more nitrogen and sulfur oxides than all of the world's cars and planes put together."

Having regained her composure, she grabbed her glass and took a swig of water.

Kasemchai smiled. "You have a great passion for your work. I truly admire that."

Liv mirrored his smile. She always enjoyed finding someone who was interested in her work, and that didn't happen often enough. "But as a business solution, it's a much more manageable project to convert every tanker ship to electricity than every car. There are only about sixty thousand ships worldwide. And it would significantly change the world for the better. We could win the Phoenix Global Water Challenge."

Kasemchai sipped from his glass of water. "Will it really change the world? What if we don't win the Phoenix Global Water Challenge? What if we run out of money first? I have a lot of money to invest, but it won't last forever."

"This is important work, Kasemchai. In truth, we don't even know if this is the right solution. We'll never know until ships adopt our solution. Maybe we will run out of money. Or maybe we will change the world."

Chef Arlo returned to their table. She held two plates, which she placed in front of Liv and Kasemchai. A mound of dark, leathery creatures were arranged on the plate. They looked like long, wrinkled worms with bright orange tips. The opposite end was black and capped with white triangular heads that looked like a cluster of teeth or fingernails.

Liv looked up at the chef. She smiled, but her eyes revealed her concern.

"For our first bite, we will start with the seashore. In Spanish, these are pronounced 'perthebes.' In English, they are called 'goose barnacles.' You can literally taste the ocean in every bite. And it's a very dangerous job to collect them. Divers on the coast of Spain must crawl out on rocks while waves crash over their backs to pluck them from the sea by hand. A few people are lost every now and then."

Liv had lifted her fork to poke at one of the goose barnacles, but on this last piece of news, she quickly placed it back down to rest the tines on the plate. "Oh, my goodness. Is it worth it?"

"I think so. After all, the best things in life do not come without risk. That makes them all the sweeter."

Kasemchai picked one up and showed it to Liv. He twisted the leathery part and pulled it off, revealing the slick inside. He took a bite, chewed several times, then closed his eyes.

Arlo smiled. "There are no spices. It is boiled in the very seawater where it lived in Spain. Then chilled in a bath of ice before rushing to your table."

"Just a hint of brine. And so soft, more like a scallop than this chewy-looking creature," Kasemchai said.

Arlo and Kasemchai both looked at Liv. She stared back at them. Then, carefully picked one up, twisted and pulled off the sleeve, and bit.

She chewed a bit more before swallowing it. "I'm shocked!" she exclaimed. "Never in a million years would I have expected this to be so delicious."

"Eet smakelijk!" Arlo bowed and walked briskly back to the kitchen.

They ate in silence for a few minutes, savoring every bite. Once, when Liv twisted the sleeve off, a squirt of brine shot across the table. Kasemchai pushed his small bread plate over the offending spot to hide it. Liv made a small "Hmm" sound.

"Do you like wine?" Kasemchai asked.

Liv chortled. Then, she clapped her hand over her mouth, realizing how loud she was. "Wine? Never. I could never afford it."

"Then I think it's time you started to appreciate it." He stared at her for a long moment. "I'll tell you a secret if you share one with me?"

She hesitated, then reluctantly said, "Okay."

"Tell me something about your business that I don't know."

Liv took a sip of her water. She cocked her head to the right, as she often did when she was thinking, and stared up to the left at the ceiling. "Well, Luuk has not been all-in on the solar panels. He's also been looking into hydrogen solutions. The problem with producing hydrogen is that it's clean but requires an incredible amount of CO_2 to produce."

"Hmm, that doesn't bother me so much. I really only care about reducing the use of fuel for ships."

Liv's face flushed. "But you can't fix just one problem. The world is more complex than that. We can't reduce the cost of shipping water and contribute to the warming of the Earth. It evaporates water. We use more energy to stay cool. Eventually, the Earth will be too warm for humans. Actually, you should think of it this way: too expensive for humans."

"All right, all right. I hear you. Now, my little secret." Kasemchai turned his glass around absentmindedly with his fingers. "Sol-vang was running out of money. If I hadn't stepped in, you never would have solved this problem."

"You're kidding."

"Nope. But now you are taking care of business. You're going to turn this around."

Liv had been spreading butter on a dark slice of whole wheat bread. She carefully placed the bread back down on the plate, and the knife on the table between them. She looked up at Kasemchai, her eyebrows knitted. "That's what I don't understand. I'm not a businessperson. Why would you make me the president?"

"Because I trust you, Liv. That's how I invest. I don't really run a business; I help set targets. I make sure we make money. You take care of me, Liv, and I'll take care of you. It's that simple."

She turned the knife over several times before trapping it under her fingers. "Can I ask you another question?"

"Sure."

"Why do you care so much about saving fuel but not about its impact on the environment?"

"It's complicated, Liv. But for me, this will get me something I need."

He leaned back in his chair and beckoned the sommelier over.

Chapter Twenty

PHILIPPE STRAIGHTENED and leaned onto the chair's arm. His back was still sore, but moving the pain from one cheek to the other felt like relief. The door opened, and he looked up as it closed behind his guest.

Preap strode towards him, crossing the vast room until he reached Philippe's desk. He finally placed his hands firmly on the surface, which was made from a single piece of African bubinga wood. The grain resembled undulating velvet, as beautiful as it was extinct.

With a gravelly voice, Preap said, "You look like shit. Don't they exercise on this side of the world?"

Philippe stood up and patted his emerging belly proudly. "If you visited more often, you'd get proper French cuisine. Then, you could learn to appreciate cheese, cream, and wine truly. Not that crap you eat in Asia."

The two men laughed heartily. With no one in the room, they put their guard down and spoke in the banter that carried these two men through years of hell. The enormous wealth they now had was a new skin for these two veterans. They walked around the desk, shook hands, and leaned in for the right arm shoulder hug.

The wall of glass behind Philippe's desk offered a view of the vast Crystal Egg atrium. At this height, a few vines hung from the ceiling, but he shared the space mostly with butterflies and colorful birds diving at each other and flying down to the canopy.

"Speaking of wine," Preap said. "I may have something you can appreciate. Ask those guys always standing outside your door to come inside."

Philippe smirked, then looked over at the door and called out.

The door opened, and a man entered. He wore a stiff-fitting outfit with a long overcoat that flared out, and gold buttons ran from his neck to his waist. He cradled a wooden box in his arm and placed it carefully on the desk in front of Philippe.

Preap pointed at the box. "Go ahead, open it."

Philippe carefully removed the lid and lifted the dark red bottle of wine out of the box. He held it up high and examined the label.

Preap beamed. "A cellar was recently discovered in southern France. This Cabernet Bordeaux predates the Rain Wall. In fact, it's nearly fifty years old."

"Très magnifique, my friend!" exclaimed Philippe. He hugged the wine to his chest and closed his eyes for a moment. "Wine of this vintage must be decanted carefully. I will save it for when you next return." He turned to the attendant. "Please, would you open a Pinot from Côte de l'Est Vinery?"

The man hurried out of the room. The door made the slightest click as it closed.

"Is it French?" Preap asked.

Philippe stretched his arms out and arched his back, then quickly straightened back up. He held one hand to the lower part of his back, rubbed it vigorously and moaned, "Aaa, why do we get old, my friend." Remembering the conversation, he laughed and, with his hand still on his back, said, "No, no, my friend. It is a play on words. The name incorporates 'côte,' which, of course, means 'coast' in French, and 'l'Est,' which means 'the east.'"

"And why is that so funny?"

"Because, my friend, it is on the East Coast of the United States. In New Jersey, to be precise.

"I can always count on you to get sloshed on the good stuff, Philippe."

"And if history is any indication, this day should be no different, Preap."

Preap recoiled in mock horror at his words, and the men laughed.

"As I was saying, this wine comes from one of the few wineries left in the United States." The two men began to walk together beside the windows looking into the atrium. Outside, the sky was grey, and they could hear the steady drone of rain on the glass panels. Inside, bright lights shone from above and illuminated the life that flitted about, enjoying another sunny day. "It is a tough business in this environment, but not impossible. I'm a sponsor. After all, the gods must make wine for the gods."

The two men laughed harder.

Philippe slapped Preap on the shoulder. "Come on, let's get down to business."

They walked over to a table with two chairs in front of the tall glass panels overlooking thick vines of philodendrons climbing up the trunks of tall palms. A cluster of orchids had been hung from one of the trees to allow its delicate pink and white petals to cascade. They seemed especially tantalizing to a group of butterflies with iridescent black and turquoise wings.

As they settled into their chairs, Preap relaxed, crossing his legs, his gaze resting for a moment on the butterflies before asking, "How's your daughter doing?"

Philippe's face softened. "She's doing very well," he replied. "The new water purification center has opened. And she has even found time for dating. I like the man very much. But the stories I hear from her. Fighting and making up. Fighting and making up."

Preap's brow lifted slightly, and he chuckled. "There are far worse problems to have in life than love, Philippe."

"There are indeed, my friend." He let out a long sigh. "Speaking of love and hate, what are you hearing about the Phoenix Global Water Challenge?"

"Lots of projects around solar and sails for tanker ships," Preap said.

Philippe's eyes sharpened. "Do they really think we'll lower the cost of water if they reduce our shipping costs?"

Preap smirked. "They have to believe in something. You know President Bob and I have a long history." He ran his fingers through his hair, and his palm lingered on the back of his neck. It felt heavy as he thought of his many experiences with the man when his name was simply Bob. "He's no dummy, but he also knows he has to pull on every string."

"He probably thinks we'll be begging for the innovation. I hear there are a lot of desalination plant proposals as well."

"Sure, but to win the award, the plant would have to be significantly cheaper or produce more water," Preap said. "Incremental innovation doesn't win awards."

"They've switched tactics. They recognize that the saline they return to the sea is making the ocean toxic. Three projects are focusing on atmospheric water generators. They're still expensive, and there are concerns about the environment when you remove that much water from the air."

Preap raised an eyebrow. His voice was raspy when he said, "That's clever. What could possibly go wrong?"

The two men sighed. These men knew what could go wrong.

A knock on the door jarred them from their reminiscing.

"Come in," called Philippe.

A man and woman entered. The woman wore a long dress of the same stiff fabric of the man's outfit. Her buttons matched the man's, but there were only three, all just below her neck. She carried two long-stemmed wine glasses with very large bowls, one in each hand.

The man opened the bottle and placed it on the table while the woman smiled pleasantly and set one in front of each of them. They bowed deeply and exited the room.

Philippe poured a tasting glass for Preap, who lifted the glass and swirled it around in the bowl, admiring its legs. He brought the bowl up to his face and inhaled deeply. His eyes fluttered for just a moment, and he smiled. He sipped and placed it back down on the table. "I'm picking up a hint of raspberry and earthy mushrooms. That's good shit . . . and it tastes expensive."

Philippe laughed and poured them each a glass. "That it is, my friend."

Preap raised his glass. "À votre santé."

Philippe raised his glass and tapped back. "Chol muy."

The men enjoyed the wine until Preap brought them back to business. "I'm still negotiating with China National Petroleum. They're not running all tankers, but they're reluctant to sell any more ships."

Philippe huffed. "They could build a desalination plant on every meter of coastline and still not produce as much water as us." He stared down into the bowl of his glass and slowly swirled the wine. "That doesn't worry me, but we need to move more water. We need to own every ship on the ocean." He looked back up at Preap, and darkness crossed his face. "Who knows how long we'll be the only source? The earth can change again."

Preap sat back and took a long sip of his wine. "We don't need *more* ships, necessarily. What we need are faster ships."

"It's too expensive," Philippe said. "You know as well as I do that faster ships burn more fuel. And I haven't seen any serious projects from the Phoenix Global Water Challenge that will make our ships any faster."

"So we're back to OPEC. What's our target for lifting the cost of water?"

"Three percent. I propose we absorb a three percent increase in bunker fuel, and then the cost of water goes up."

The two men raised their glasses and toasted. It was one of those perfect moments where the crisp sound hung and fell in repeated waves, gracefully fading away.

"And how's your negotiation with the Saudis going?" asked Philippe.

"As you would expect." Preap almost closed his eyes, and then they sprang back to life. "They are beginning to buckle, but they are not the only member of OPEC."

Philippe frowned. "Why do we even care about OPEC? There's so much oil from China and the US. Even Brazil."

"You know why. Their rate influences the world's rate. They lower supply when demand falls."

"And we raise the price of water." Philippe waved his hand dismissively. "They're stubborn."

Preap raised his glass in the air. He tilted the glass just enough to punctuate his point but not so much as to spill the ruby liquid. "And it's hard work. I wouldn't want to oversee a crew wearing cooling suits in that furnace. I'll take water over oil any day."

"You're too forgiving, my friend."

Preap smirked. "Those are fighting words, Philippe."

The two men sat back and reflected.

Philippe stared up at the ceiling where an enormous chandelier hung. It didn't really hang so much as emerge from the ceiling in the form of the base of a knurled tree. Where the buttress roots flared out and met the imaginary ground, strings of light dripped into the cavernous room. "I forgot to ask, Preap. How's your son?"

"Oddly enough, he's funding some company building solar-powered container ships. Who knows, maybe there's potential. Something different from all these dating PrefXs." He took a long sip of wine and smiled broadly. His teeth slightly captured a shade of pinot. "He thinks he's a self-made man. But whenever he wants serious funding, he goes to an associate of mine in Japan. We'll keep him on the right path."

Philippe laughed. "If only our children knew how grateful they should be. They are destined to follow in our footsteps. It's probably time to start teaching them how the *real* world works."

"Are you saying they are ready for the roster? I don't know." Preap turned his head. A mischievous grin crossed his face as he turned his eyes back to Philippe. "It's like teaching your children about sex. I enjoy it. I know they will enjoy it. But telling them about it takes all the pleasure away."

Philippe was just about to drink from his glass and struggled to return the glass to the table as the men roared with laughter.

Once he had regained his composure, he placed the disc from his AugWatch on the table. "All right, all right, let's bring up the roster—time for fun. Let's focus on the United States first."

He swiped his finger over the disc and flicked a huge digital board before them. It showed the US. Many of the states in the center of the US were grey. Those were dead states. The only states with people now were those on the coast that could receive water shipments from their ships. Each of the states that was not grey had two heads, each a US senator.

"I'd like to focus on the Democrats," Philippe said. The heads of the

Democrats all became larger. "I like their message on fiscal responsibility and family values."

Preap was about to take a drink and lowered his glass. "I thought we agreed we were going to flip the teams again. We need more water infrastructure. The Democrats need to embrace federal projects again."

"I'm just giving you a hard time, my friend. As you like to say, 'It's good business to make bad politics.'"

Preap studied the chart. "You had me worried. Okay, right to left, starting in Maine. We remove Senator Duncan. He's pro-desalination."

Philippe spoke to his companion, Mélanie. "Destroy Senator Duncan's career, please."

"It is done, Philippe," it responded. "Senator Duncan often gives his two children a bath on Sunday. A very wasteful use of water."

Philippe looked up. "Uh, is that enough, Mélanie?"

"He is also the sponsor of the Weekend Water Waster bill. It increased the cost of water during the weekend. This was largely seen as a tax on people living in high-density housing because they have smaller water tanks and generally buy water daily."

Philippe grinned. "So, you're saying this senator is not only using an exorbitant amount of water but he's also created a tax so that less fortunate citizens can't afford to do what he does freely?"

"You've got it, Philippe. He'll deny it and claim it's the Republicans trying to slander him. I have a plan for that. It will make his fall even more spectacular."

Philippe looked up in surprise. "Are you enjoying this, Mélanie?"

"I don't have emotion, Philippe. What you detect is my mannerisms, which are modeled after yours."

"Touché!"

Preap chuckled and raised his glass. "To fair competition and the world's never-ending thirst for a good contest."

Philippe's smile widened as he raised his glass to meet Preap's. "Let's give them a contest to remember. After all, what's a game without its worthy adversaries?"

Their laughter echoed softly, each man acutely aware that they held the reins. While they often acted as rivals, it was their shared stage—and they intended to keep it that way.

Chapter Twenty-One

LIV SLID her finger across the dusty counter, revealing its true color. The hiss of steam stopped, and the automated tea sommelier spun around and moved Liv's tea towards her. She flung her thin scarf across her face and grabbed the cup. Outside, the sun was hidden behind a sky that seemed slightly less amber-colored than the previous day. The diffused rays didn't sizzle her skin, but at one-hundred-and-nineteen degrees Fahrenheit, it still felt hot.

She sprinted across the road and into the building lobby. Once inside, she pulled back the scarf and unconsciously took a deep breath. She took a sip of her hot tea and immediately regretted it. "Ah! Hot."

"Did you burn yourself again?"

Liv could swear she detected a tone of mirth in her companion's voice, but quickly checked her imagination. AI couldn't have emotions.

"Yes. You know me so well, Noa."

Only then did she realize her AugWatch was gently buzzing on her wrist.

"Chantal is calling you, Liv," Noa said.

"Good morning, Chantal. What's up?"

"I'm getting the conference room ready for your meeting with Cape Town at seven."

"Good. I saw the revised contract this morning. Could you ask Emma to join us? I want legal to sit in on this one."

"Got it. See you upstairs."

She continued across the lobby, and when the elevator door opened, she stepped in. She turned, watching the doors as they cut her off from the outside world. She shivered uncontrollably.

"Eighty AI collaborators," she said out loud. "I can't believe I'm hiring eighty AI collaborators in another country. This is crazy, Liv."

She watched the numbers above as the elevator rose. They had expanded their office space, and she passed her old floor and continued up seven more. The doors opened, and she saw Hendrik and Chantal waiting for her.

"Good morning, Hendrik. Good morning, Chantal."

Her team responded in near unison, "Good morning. Are you thirsty?"

"No, I'm not thirsty."

Chantal started to raise her hand and dropped it. "I need to talk to you before your seven o'clock, Liv. They left off the security screening details in their last proposal."

"I know," Liv assured her. "It was in the previous proposal. I'm sure it's a mistake."

"Kasemchai said he wants a check-in next week," Hendrik said.

"Is he coming here or virtual?" asked Liv.

"This one's virtual."

"Good. Set it before his lunchtime."

"Got it."

The three stood for a moment longer. Hendrik rocked on his heels until he caught a side glance from Chantal. He grinned and said, "Uh, maybe you'd like to go to your office, Liv?"

Her cheeks flashed red and she stammered, "What? Ah, I . . . I forgot." She turned to walk to the door and turned back just as quickly. "Hendrik, could you go to my office? I mean, could I see you in my office?"

"Of course."

People paused their conversations or looked up from their chairs and waved to her as she passed. Liv waved and said good morning to each person as she walked towards her office. Inside she dropped her bag on the floor beside her desk and paced in front of the window, blowing on her tea.

"Could you close the door?" Liv asked.

Hendrik closed the door behind himself. He was easily a foot and a half taller than her.

Liv clasped her cup in two hands and blew the steam over the top with a long exhale. She looked directly at Hendrik, and her eyes shot open wide. "What the hell am I doing, Hendrik?"

"What do you mean?"

"We're hiring eighty people today in South Africa. I don't even know their names. We've opened an office in Germany and two in China. Our budget for AI modeling is greater than this company has invested in the past seven years, and I'm working on doubling it right now."

"Yes, I know."

"This is everything I wanted. We have all these minds working on the PV panels and the batteries. I . . ." She stopped. The last word seemed to be stuck in her open mouth. She stared through Hendrik, not at him.

Hendrik pursed his lips and nodded. "Don't worry, Liv. You're doing great. You hired them. Yeah, it's a lot of people, but we need many people. You can't focus on every person. That's what you have managers for. *You* need to focus on where we're going, captain."

Liv blushed. She hated it when he called her captain. It made her feel too important, but she also loved it. It reminded her of a simpler time on Ishigaki Jima, and that helped her relax. "Thanks, Hendrik. All right, let me check some more messages, and then we'll get started."

Hendrik smiled warmly and exited her office. Pushing the door open, he turned and called back to Liv, "One more thing. Luuk said he wants to see the specs on the new panels."

"Shit, I completely forgot. I'll pull them up now. Too much thinking about the big things, Hendrik. I'm forgetting the important details."

"Don't worry about it. You can't do it all, Liv."

"Yeah. Yeah, of course," she replied, nervously rubbing her red and sensitive knuckles with her left hand. The nervous habit was a painful reminder that being important was not enjoyable.

Chapter Twenty-Two

IN WINE-MAKING, THE TERM "CRUSH" traditionally referred to the period between August and November when grapes were harvested and crushed into juice. However, at the Newark Winery, smaller batches were processed six times a year, making the crush period more frequent. During this period, everyone worked long hours, and the sweet aroma of grapes being pressed and the sound of machines transforming the fruit into juice filled the air.

A line of employees, covered head to toe in white clothing, moved across the floor toward the door. Makoa and his colleagues joked about the last-minute gathering as they filed into the break room for a meeting with Emmet, the winery's owner.

Emmet was a jittery man by nature. Anyone who ever met him quickly understood why. It wasn't nervousness; his brain was always running at full speed. It was all he could do to slow down to the pace of a normal human and share what was on his mind. He could jump from chemistry to finance to an observation about a bug on the way to the office with equal passion and detail.

He wasted no time in bringing the meeting to order. "I've got some exciting news, folks. Vertical farming is a fast-paced business, and we must always stay on the cusp of innovation. There is no doubt that wine is an extreme luxury, and our customers pay dearly for it. But it is also a capital-intensive business."

Emmet looked out at his staff. He smiled and shook his head twice —one of the signs that he was mentally backing up to slow down his conversation for his audience. "An investor recently approached us. Now, let me tell you that I took this with the same critical approach that I use for all business decisions. And top of mind in this decision was the strength of our staff."

Makoa nodded in affirmation but wished Emmet would get to the point instead of emphasizing the importance of staff. Talking about business was always good. Talking about employees always made him worry there'd be more work or fewer people.

"I suppose I could have waited until the end of the crush, but this is just too exciting. I want to let you all know that we have a new partner." He shoved his hands deep into his pockets. "I think you all know the Seng family."

He paused to let them laugh. Everyone nodded enthusiastically. It was an odd question, like asking if everyone knew the sun. Everyone did. Not personally. Not up close, but there was no life without the Seng family or the Labonne family.

"The Seng family." He pulled his right hand from his pocket and, with his hands slightly curved, gave the air a soft chop. "Specifically, Kasemchai Seng has taken an interest in our winery. He is now a forty-nine percent owner of the Newark Winery, which includes significant financial commitments. His generous investment will allow us to increase our real estate and purchase new equipment. This will allow for a significant increase in productivity." He held two thumbs up. "Now, this doesn't mean we will sacrifice quality for quantity. The market for wine has not grown dramatically, but the thirst for innovation has. We will be exploring new varietals, increasing the size of our aging cavern, and even expanding our work with ultrasonic waves and gamma irradiation to speed up our aging process."

The news was met with a mix of cheers and curious glances, but Makoa's expression darkened. The winery had always relied on water from Africa. Would this mean switching to Asian water?

It wasn't like Kasemchai or the Seng family were evil, but he knew where Philippe stood. He stood with God. Bringing in Kasemchai was a betrayal of the values the winery held dear.

The staff returned to work. Makoa lifted crates of grapes from an endless stack and poured them into the hopper at the top of the crusher destemmer machine. A pair of rollers broke the skins open, separating fruit pulp on one side and stems on the other. The machine was quite noisy, leaving Makoa alone with his thoughts while the machine did most of the work.

He jumped when Emmet tapped him on the shoulder. He grinned sheepishly and leaned over to turn the machine off. The silence was immediate.

"Makoa, I wanted to check in with you. I noticed you didn't seem as elated as the others with the news. What's on your mind?"

"Oh, it's nothing," Makoa said. "I'm sure it's all good news. I was just wondering how things would change."

"Well, for the better, I hope." Emmet stared at Makoa for an uncomfortably long moment. "A little extra funding will be good for everyone's salaries. But I don't think that's what's on your mind."

Makoa stared at his feet. The toes of his white rubber boots were stained with overlapping smears from the red grapes. "Look, it's God's will that the water falls in two places. We have to work with what God gave us, but that doesn't mean we have to work with the Asian water family. The Sengs do not have God in their heart."

Emmet looked up with a stern look, and then his face softened. "Look, it's not because of God that the rain only falls in two places. It's because of humans. It's because of the decisions we made. Or we didn't make."

"I can't believe that, Emmet," Makoa said. He could feel his temper rising. The tips of his ears began to burn. "If humans could have made a difference, then we would have done something about it. The world would be different." He looked down and shook his fist at the floor. "Hawaii wouldn't be barren." Makoa shook his head and paced back and forth. "God gave us this Earth. No human could ever believe that their needs are greater than God's. That their needs are too important to respect what he has given us."

"We did. We prayed to the God of man. Man is greater than God, and no one can tell him what to do," Emmet said.

Makoa's mind raced. This conversation was making him furious, but he didn't want to lose his job. He needed to bring it back under control. "You shall not make yourself an idol. God is a jealous God."

Emmet let out a long sigh. "I'm sorry. I didn't mean to make this about God. It's just, it's just business. We need to keep the winery in business."

"Is business the only decision we have to make, Emmet? I know this is good news for the winery. But I can't help but think about an investor like him. It's good to grow the business, but they . . ." He shook his index finger in the air. "They do not believe in God."

Emmet placed his hands into the front pockets of his white lab coat. "Makoa, we don't know what Mr. Seng believes in. And I understand that change can be daunting. I'll keep a close eye on how this partnership evolves, and I can assure you that our values will remain the same while we adapt to changing times and explore new opportunities."

"Sure. Sure," Makoa said. He wasn't sure at all. But he kept that to himself.

Chapter Twenty-Three

KASEMCHAI STEPPED out of his car and onto the very narrow sidewalk in Kyoto. He stopped to admire the sidewalk's stone edging, straightened his starched-white shirt, and turned back around. He faced a doorway obscured with three light blue cloth panels. In the center was a large Japanese character that looked almost like an angelfish.

As he parted the curtains and entered the room, his eyes met with the warm smile of a woman sitting behind a desk.

"Seng-san. Please, you can undress in there." She pointed to another door obscured by three curtains. "Sato-san is waiting for you in the bath."

Kasemchai pulled the curtains apart, bit his lower lip, and walked through. Inside were several small benches and rows of lockers with keys sticking out. He removed his clothes and placed them into the tiny locker.

Through a steamy glass door, Kasemchai could barely make out Mr. Sato sitting on a small blue stool. Covering his privates with his right hand, Kasemchai pulled back the sliding door with his left and entered the steamy sentō.

"Hello, Mr. Sato!" he shouted across the room, a little louder than he intended. His voice echoed off the walls.

Mr. Sato's head snapped up. He grunted, and with the help of the hand on his knee, he stood to face Kasemchai. Sudsy bubbles dripped down from his shoulders. The shaving cream on his face clearly indicated the job was half finished.

He bowed. "Hello, Kasem-chan. I'm so glad you could meet me here."

The two men bowed several times. Kasemchai tried to bow just a little lower each time.

"Of course, Mr. Sato. I remember this sentō well."

As Mr. Sato sat back down on his stool, he motioned for Kasemchai to join him on a nearby blue stool. Kasemchai sat and gazed at his reflection in the mirror. On the ledge below it, there were several small bottles holding shampoo and soap.

Kasemchai reached for the silver faucet in front of him, adjusted the hot and cold knobs until the water was hot but not scalding, and splashed it on himself. "I have to admit, I still feel a little silly in here, but I remember how relaxed and clean I always feel afterward."

Mr. Sato smiled. He turned back to the small ledge in front of him, grabbed a bristle brush, and applied shaving cream to his jaw. "You know, Kasem-chan, there is actually hidden meaning in the sentō. For example, do you see those two faucets in front of you?"

"I do."

"We call it a 'ka-ran.' It's a Dutch word that means 'faucet.'"

"Oh, I didn't know that."

"Do you see the long neck of the ka-ran? Can you guess the other meaning this word has in Dutch?

"No idea, Mr. Sato."

"It's the Dutch word for a bird—the crane."

Kasemchai smiled and stared at the neck of the faucet. "Fascinating. The hidden story of words."

"Really, it's a hidden history, Kasemchai. In the early seventeenth century, the shogunate feared threats to his power, especially in the form of Christianity. He imposed a period of self-isolation to protect Japan from foreign influence. The only Western country allowed to trade were the Dutch. And only on Deshima, a small island surrounded by walls."

Kasemchai feigned surprise and wonder. He knew a lesson was

coming but couldn't imagine what it possibly could be. "How long did this last?"

Mr. Sato pulled down on his right cheek and drew a silver-handled razor down his face, leaving a clear path in the foamy lather. "More than two hundred years. But there are consequences when you ban something. The Dutch introduced Western medicine, but high-ranking officials went to the island to receive modern medical treatment. The Dutch language became the language of science." He turned and shook the razor toward Kasemchai, the right side of his face mostly clear of shaving cream. "But there were a thousand holes in this policy, and eventually it collapsed."

"I don't understand. Is this a story about language?"

"In a way. Japan wouldn't be the same without foreign ideas. But more importantly, Kasem-chan, remember that you can't solve complex problems with simple solutions."

Kasemchai slowly nodded. "Uh-huh. I see." But the meaning remained a mystery to him.

Mr. Sato had already turned back to the faucet and was splashing water on himself to rinse off the remaining shaving cream.

Kasemchai smiled and turned to face his mirror. He lathered his head and body. Then, he turned on the water and, using a small bucket with a long handle, rinsed off the lather with water from the tap.

"The Serendipity PrefX is doing well?" asked Mr. Sato.

"Two-point-three million new users last month. China is our fastest-growing region right now."

"That's good. And where is your next opportunity?"

Kasemchai smiled. "You're always one step ahead of me, Mr. Sato."

Mr. Sato laughed. "That is my superpower. Your power is quick decisions. Mine is predicting human nature." Mr. Sato poured a final bucket of water over his head. "Are you ready for a soak?"

"Of course."

The two men walked towards the elevated tubs behind them. There were four raised tubs, all beautifully decorated in tile. Mr. Sato pointed to the first one, which vigorously bubbled. "That one is hot with jets. That one has minerals from the silver spring in Arima Onsen—the oldest hot spring in Kansai. That one is warm with no bubbles, and that one is cold. Where would you like to start?"

"Let's start with the bubbles."

"Very good," Mr. Sato said. He walked over and lowered his body into the water.

Kasemchai averted his eyes so he didn't have to watch the older man. He always found the bath a mixture of relaxation and confusion, surrounded by naked older men who frequently sighed loudly and grunted. Kasemchai lowered himself into the same tub with ease. He didn't look forward to his body aging, but he did envy their confidence.

Mr. Sato regarded Kasemchai from across the large bubbling tub. "Tell me, Kasem-chan, what is next?"

"Well, I'm looking into energy."

"Ehh?"

"Solar energy, to be exact."

"There is no shortage of sun these days, Kasem-chan. And no shortage of competitors either."

"Yes, but I have a niche—tanker ships. Like the ones that bring your water to Japan."

Mr. Sato let out a long sigh. "So desu neh." A popular Japanese phrase, it was easily recognizable as "I see." It was neither confirmation nor rejection. A confusing statement that made Kasemchai's mind race.

Mr. Sato got up from the jet pool and entered the mineral pool. Kasemchai followed.

"Why solar power?" Mr. Sato asked.

Kasemchai wiped the perspiration from his forehead with the back of his hand. "Tanker ships run on the dirtiest fuel. It's called bunker fuel. It's dirt cheap, but prices rise erratically. I often hear Father complaining about the Saudis and their grip over his business."

"Yes, even cheap oil in enough quantity is a heavy cost."

"Water is a necessity, not a luxury. By adding solar panels to the ships, they could reduce their fuel consumption. Even a little would help."

"Ahhhh," Mr. Sato said. He pinched his lips and rocked his head in affirmation.

Kasemchai waved away the tiny bubbles that clustered around his chest. "Did you know that just a handful of the biggest tanker ships produce more nitrogen and sulfur oxides than all of the world's cars and planes put together?"

"Ah ha," Mr. Sato said. "You are a businessman and an environmentalist."

"Business first, of course, Mr. Sato," Kasemchai smiled. It was impossible to know if he approved of his venture.

Mr. Sato sat back. Every few minutes, he would say, "Umm," "So desu," or "Uhhhh."

Kasemchai was used to Mr. Sato's thought process, but it didn't make it easier for him to hold back the conversation. The pauses were not meant to be an invitation for Kasemchai to speak.

Finally, Mr. Sato said, "I see." Then he got up and moved to the still pool. "Kasem-chan, I like your idea very much. I know both your father and Mr. Labonne would like to reduce their reliance on fuel. It is a great business for the world."

"Thank you, Mr. Sato. When I heard of this idea, I thought, 'This is a noble cause.'"

"Yes. A calling, if you will." Mr. Sato smiled and gazed at the ceiling. He seemed to be dreaming of the possibilities when, without warning, he heaved himself out of the tub and walked over to the cold tub. He gripped the rail tightly and lowered himself into the frigid water, then heaved himself out. He repeated the process, each time lowering himself deeper into the icy water. Each time he lowered himself, he squeezed his eyes shut, clenched his jaw and hissed, "Ya ta."

It was a sharp statement. The "ta" was smothered by a long exhale.

Kasemchai trailed him into the cold tub. He hated this tub. It stung and seemed to contradict everything else in the sentō. But he also felt a rush of energy, and his chest felt clear, like he was breathing in mentholated air.

Mr. Sato shook his head. Then, he considered Kasemchai. "You have created the greatest love PrefX in the world. You are successful there. It seems your calling is there. Solar-powered ships sound like the kind of pursuit certain wealthy people make when they are trying to decide what to do with all their money. "

Kasemchai shivered as goosebumps raced from his toes to his scalp.

It was not from the water. "You're right, Mr. Sato. I have been very successful with the Serendipity PrefX, and I recognize your contributions to that success." Kasemchai tried to pause. He wanted to sound measured. To show he was a great thinker. But the cold water made him shiver uncontrollably. He needed to get out of the cold pool. "This is a big business. Ships move water and goods. We keep saying we'll reduce our use of oil, and every year, we find new ways to increase our consumption. Where there is volume, there is room for disruption."

"Disruption. Yes, still a popular term for unbridled change," Mr. Sato said. He closed his eyes and looked down. "Yata!" Then he grabbed the silver railing and vigorously pulled his body out from the frigid water. He moved to the jet pool again and lowered his body, grimacing in pain.

Kasemchai followed him. The hot water stung his frozen skin like a thousand pins, and it seemed to burn. As he submerged his body further, he felt it embrace the warmth.

Mr. Sato said nothing.

Kasemchai waited. His eyes darted between the ceiling and the water.

"Is this solution for both water families?" asked Mr. Sato.

"Of course. I hear them both complain constantly about the cost of fuel."

"Does it make you their equal?"

Kasemchai rolled his eyes upward. He thought about this. "Not their equal, of course. But important."

Mr. Sato said nothing. He closed his eyes. Possibly thinking, possibly sleeping. It was hard to tell. He opened his eyes. "Why do you need to be important?"

Kasemchai rushed his answer. "I don't need to be important. I can help them with their negotiations with the Saudis. They will want me to be there."

Mr. Sato thought about it. He absentmindedly ran his tongue over his eye tooth, making a sucking noise out of the side of his mouth. "I think you have already made your decision to pursue this business, Kasem-chan. When you have an idea, you follow it relentlessly. That is something I can invest in."

Kasemchai smiled. Unconsciously, he dipped his head a bit. He wanted to say something but waited.

"You said you liked disruption. Well, let's make it worth our time." Mr. Sato sank deeper into the water until it lapped against his chin. "Solar panels are the beginning of an idea. We need a bigger idea. Fuel efficiency is not the solution. If oil is the problem, then get rid of oil. Kill it. Change the world forever."

Kasemchai felt queasy. He had been hoping for help. He had been expecting a challenge. He hadn't expected to be asked to change the world. He could feel the sweat on his skin. He wondered how that was possible in a hot tub.

Mr. Sato chuckled. "That wasn't the answer you expected, I can see." He got up and moved to the mineral pool, Kasemchai trailing behind him. "I'm certain you already have a good team working on this. Your job is to help people see what is in front of their faces, but they don't see. Give them a challenge. Give them a deadline. They will find the solution you are looking for."

Kasemchai let out a loud sigh. "Okay, I think I can do that." He looked down and then back up. "You will invest if we do this?"

"If you present an idea to me that changes the world, you will get my backing and the exposure that you need. And one more thing."

"Yes?"

"I know your group is based in Europe, but remember, ships are made in Asia." Mr. Sato stood and pulled himself out of the hot tub. "Good luck, Kasem-chan. I'm hungry now."

He walked towards the sliding door and into the dressing room.

Kasemchai stared after him. He wondered if that was an invitation to join him for dinner. Or goodbye. His mind wandered. What was his next move?

Chapter Twenty-Four

SEVERAL WEEKS LATER, Emmet walked between the tall maze of walls with Kasemchai. He pointed to the thick leaves of the grape vines and explained how they grew horizontally along the wires. He dipped his finger into the soil and explained how they managed the terroir to develop regional-specific wines.

The tall parallel guide wires of a robot gardener moved towards them, and Emmet pointed out how it tested each grape cluster and plucked them at precisely the right moment.

Kasemchai nodded appreciatively. He pointed at things that stood out and didn't look like the tour bored him in the least. They rounded one wall and started down the next aisle when they came up on Makoa, who was monitoring the work of a robot.

Emmet observed Makoa for a moment while he worked, then caught his eye the moment he realized they were there.

"Makoa, I'd like to introduce you to Mr. Seng, our new partner."

"It's nice to meet you, sir. Are you thirsty?" Makoa said.

Kasemchai pressed his hands together in front of his chest and bowed slightly. "No, I'm not thirsty."

Makoa looked around, unsure of what to do, then bowed awkwardly. He'd never had to bow to someone before, but then again, he'd never met someone from the Seng family. It was impossible not to feel in awe of him.

Kasemchai grinned, but it was a kind look. "Tell me, what are you doing here?"

"Basically, I'm checking the robot's work. They're very accurate, but from time to time, they make a mistake. When I see it's made a mistake, I tell it what it did wrong so it can learn from the experience."

"These are really amazing robots," Kasemchai said. "But I have to say, it's the soil that really blows my mind. I can't believe you can match the actual terroir of a French Bordeaux, a German Riesling"—he waved his hand back and forth at the growing tower—"or a California Pinot."

Makoa brightened at the conversation. "What's most important is we don't have to make a choice between a Reisling or a Pinot for a plot of land. Every varietal gets exactly the soil composition, even temperature, and compactness for the wine style we wish to achieve."

"It's amazing. And the truth is, I love to share. I didn't invest in the Newark Winery to keep it for myself. I want all my friends to enjoy your finest."

Emmet laughed. "Or at least our fellow oenophiles."

"That's right," Kasemchai said. "Quality is of the utmost importance. But that doesn't mean it can't be the most sought-after wine in all of Asia."

A darkness crossed Makoa's face.

Kasemchai continued, "My family has a long history of supporting businesses that aim to make the world a better place. My father always thought being benevolent was one of the greatest guiding principles in life."

"I'm sorry. What does *benevolent* mean?" Makoa asked.

"Well, it means to genuinely want to do good, to be kind and generous in our actions, especially towards projects that have a positive impact on the world." Kasemchai gazed at the rows of vines that towered over him like a pilgrim in a sacred temple. "Supporting the arts is just as important as bringing water to the world. And your wine is truly art."

"I don't mean to sound negative, but not everything is as it seems in the world of business. I'm sure it may seem benevolent to ship water or wine, but isn't it just business at the end of the day?"

Emmet stepped in. "Thank you for telling us about your work, Makoa." He turned to Kasemchai. "Let's head down to the area where the grapes are turned into wine."

Kasemchai held up a hand to pause him. "I appreciate your candor. Sorry, what did you say your name was?"

"It's Makoa."

"Thank you. As I was saying, I appreciate your candor, Makoa. I've heard my fair share of cautionary tales, but my family genuinely wants to make a positive impact. I understand your skepticism, but I assure you, we're doing good work."

Makoa looked down at his feet for a moment and then stared Kasemchai in the eyes. "We have all been through some challenging years. You have to understand I feel very passionate about the wine we make here. When they say wine is the blood of Christ, of course, no one thinks of it as really his blood. But it brings you closer to him—His divinity."

"And that is why I invested in the Newark Winery." Kasemchai placed his right hand on his chest. "The wine stands on its own merit. Your technology is second to none. And passion—that can't be imitated."

He reached out and bumped Makoa's fist. Emmet laughed awkwardly, then steered Kasemchai down the aisle as quickly as he could.

Chapter Twenty-Five

Back in Sihanoukville, Kasemchai ran his hand over the enormous iron sculpture of three manta rays he had installed in his office. Its back was smooth and cold to the touch. He looked at his AugWatch and saw it was 1:00 p.m.

"Chanthavy," Kasemchai asked, "what time is it in Rotterdam?"

"It's 8:00 a.m., Kasemchai."

"Good," he said to himself. "She's probably in the office."

He made his way to the bar, grabbed a glass of water, and slumped into a chair behind his desk. The chair resembled a large, off-white pumpkin with a sunken top for him to sit on. A low table separated his chair from two chairs opposite from his. They were similar but less cushioned and more upright.

He placed his AugWatch disc on the table before him.

"Chanthavy, is Liv available?" he asked.

"I'll check for you, Kasemchai."

Liv's avatar appeared before him. It sat in the chair across from his desk. "Hey, Kasemchai, how's it going?"

"I'm well, Liv. Are you still riding your bike to work?"

"Yes. You know it's very healthy to ride a bike. And it helps me think in the morning."

"No, no. I totally agree. But please tell me you at least moved closer to the office, as you promised."

Liv's avatar sighed loudly. "Yes, of course, Kasemchai. I'm in that big, swanky apartment you told me about, just like you asked."

Kasemchai chuckled, raising an eyebrow. "Good. I want you to be happy."

"Speaking of moving. The team of AI collaborators in Cape Town is making great progress. They have a materials lab that can create some really neat components for testing."

"Excellent. That's a great move, Liv. More bodies, more minds."

"Of course." Liv turned her head and looked off into the distance, gathering her thoughts. "It still makes me so nervous to have so many people working for me. I can't possibly see all their work."

"That's not the point, Liv. Your job is not managing people. It's results. We need as many minds as possible thinking about the problems. The solutions will come."

Liv's avatar visibly relaxed in the chair. "Thanks. I know you're right."

"I don't need to be right, Liv. I need, as they say in the US, to back the right horse." Kasemchai stood up. He walked behind his chair and placed both hands on the back. "It's time to talk about funding, Liv."

Her eyes darted to his. "What do you mean, funding? I thought you were funding our company?"

"Of course, I am—the lion's share. But funding is not just about money. It's also about validation. When it comes time to pitch our business to customers, they will ask for our solution because they'll know about all the heavy hitters backing us. Sol-vang will be in the news every day."

Liv shrugged, "Okay. Isn't that something you do? What do you need from me?"

"Don't worry, Liv. That's my responsibility. What I need from you is a bigger business plan."

Liv stared down at her hands in her lap. In the awkward silence, Kasemchai began to speak, but she cut him off. "I don't understand. I already gave you our business plan. You want more details?"

"No, Liv. Bigger. More incredible. A bigger idea."

"I still don't understand. The model is quite simple. We create solar panels that are added to ships. They reduce the demand on fuel, which in turn allows them to run faster."

Kasemchai leaned forward towards the AugWatch disc. "Hmmm. Yes, that makes sense. It makes sense to you and me." He stood up straight behind the chair. "The thing is . . ." He raised his hands, showman style. "To get investors, we must sell a vision—a grand idea. People don't invest in common sense. No one wants a ten percent return. Or even a twenty percent return. What they want, Liv, is a one hundred percent return."

Liv didn't rise to his excitement. "I still don't get it, Kasemchai. What does that have to do with our business plan? I can't project what kind of return people will get. Why don't you just ask your father for more money?"

"I cannot ask him for money!" Kasemchai barked. He stopped and gripped the back of his chair, his knuckles drained of color. He stared at his feet for a long moment before inhaling forcefully through his nostrils. "Sorry, Liv. I didn't mean to yell. What I need you to do, Liv, is paint a grander vision. Don't just tell me that people will buy our product. Tell me how this will change the shipping world forever. How can you make tanker ships faster? How can we move more products?"

He walked around and fell back into his chair. It gently caught him with a *phoomp* sound. "There was a moment in time when we realized that horses would no longer pull wagons. That the combustion engine was stronger and tireless. We need that moment for solar power. A moment when civilization will realize that the combustion engine is weak, dirty, and ancient. Do you know why we measure engines in horsepower, Liv?"

Liz flickered her eyes and breathed in deeply through her nostrils. A signal that she knew this history far better than Kasemchai. "Of course. They wanted to show the strength of an engine in terms that people could understand at the time: one horsepower, five horsepower, etcetera."

Kasemchai was waiting for that answer. She may know science, but he knew people. "No, Liv—to show how weak the horse was. An engine that could replace five horses was inconceivable. Now, we measure engines in hundreds and tens of thousands of horsepower. It is irrefutable that we had to switch to the combustion engine. How can we show that in a solar-powered ship? How could we replace one of those twenty-foot-tall carbon-spewing electric generators? How many combustions can we replace?"

"That's not a thing, Kasemchai."

"What's not a thing?"

"The engine isn't a 'combustion.' That's how the engine works. And it's an internal combustion engine. No one wants engines to burst in flames."

Now it was Kasemchai's turn to sigh in frustration. "I know that. That's not the point, Liv. You need to come up with a name for those old engines. And tell the world how they will disappear. Show us how old men will sit around ponds and look at tiny models of tanker ships with tiny streams of black soot smoking out the back. They'll mumble facts about ship types and grades of crude oil. They'll wait patiently for schoolchildren so they can tell them a story of the romantic days when oil was king."

Kasemchai stopped to compose himself. He looked out the window for a moment and then back at Liv's avatar. "Can you do that, Liv?"

"I . . . I guess."

"No, Liv. I need you to speak with the same passion as when I met you in Ishigaki Jima. I need that Liv. The one who absolutely knew that she would change the world. This is your moment."

"Okay. Okay, I can do this. I'll pull together the team." She took a deep breath. "We'll do some brainstorming. Maybe a retreat. When do you need this?"

Kasemchai stared at Liv's avatar. It was a little annoying that they never rose to the same level of passion as the person they mimicked. "Three weeks."

Liv's avatar gaped at him. "Are you out of your mind?"

Chapter Twenty-Six

LIV PACED BACK and forth in her office. She walked to the window and stared out across the land toward the direction of the ocean. She couldn't see it from here—it was too far off—but she knew it was there. The ripple marks in the dusty riverbed beside her office traced the path where rainwater once rushed toward the sea.

"What are we going to do, Liv?" she said out loud. "You have the team in Cape Town. New teams coming online in China and Germany. At least you can hop on a train to meet the team in Germany."

In a mocking voice imitating Kasemchai, she said, "You need to come up with a big plan—something that will change the world."

She picked up one of the model ships in her office, turned it upside down, and shook it. The little bits of solar panels and batteries and engines fell onto the desk. "Just turn the whole damn business upside down. How hard could that be?" She stared at the pieces scattered across the hard surface, then she turned and strode to the door. She flung it open and shouted out to the office floor, "Helena, Johannes, and Hendrik, could you come in here? Chantal, you're taking the meeting with Cape Town."

Chantal looked up, her eyes bright white, "Wha . . . what do you mean I'm taking the call? You're not joining me?"

"You know what to do, Chantal. Review their results. Find out the challenges, and tell them to work faster."

Helena, Johannes, and Hendrik filed into her office, and each sat in a chair. Liv closed the door, stomped to the window, turned, and leaned on her hands against the ledge. "I just got off a call with Kasemchai. He said he wants to pitch our business to more investors."

Johannes sniggered. "I thought Kasemchai was our gracious benefactor?"

"It's not for the money. It's for credibility. He says customers will take us more seriously if we have serious investors."

Helena slapped her hand on the armrest of her chair. "So what does he want, then?"

Liv pushed herself off the ledge and strode into the center of the room. "He says our business plan is solid. But to bring in the big investors, we need something extraordinary. A vision of how our solar arrays will not just work, but change the shipping industry."

Johannes looked skeptical. He glanced down at the pieces of the ship strewn across Liv's desk and then back up to her. "So how do we write a really big plan about something we don't know? We're an engineering company. This is not a dating PrefX, although I'm not sure I would say his PrefX has changed the world."

"He has created millions of relationships with his dating PrefX," Helena snapped back, "and I think the success of his business gives him the right to tell us how to woo investors."

"Sure, how hard can it be to invest in someone else's company when you're already a multi-billionaire?"

"What are you complaining about? He's investing in our company."

Liv raised her hands in the air. "Woah, woah, woah. What's going on here? We need to work, not bicker like children."

Johannes scowled down at the floor. "Sorry, Liv. You're right. What do we need to do?"

"Thank you. As I said earlier, we need to think beyond an incremental boost from solar. What will happen when all the engines are gone? What will happen to transportation? How will this impact the environment?"

Helena slouched in the chair and stretched her legs out in front of her. "What other industries could we expand to?"

"Exactly, Helena."

Johannes stood up and began to pace. "All right, if we're going to think big, we need to think about all the ways solar could change the world, starting with the shipping industry. How will this change regulations, carbon incentives, new streams of revenue?"

Helena stood up. "We're good at this—it's time for research. We'll need to do some studies, but we have plenty of funds for this kind of work now."

Johannes stood in front of a window and looked out over the city lights. He turned back with a twinkle in his eye. "That's right, Helena. Let the data be our light. We've never had the luxury of thinking beyond the simple goal of reducing fuel consumption."

"There's an energy symposium in Geneva in November," Helena said. "That gives us months to prepare and show up ready to ask questions."

Johannes stopped pacing. "How much time do we have, Liv?"

Liv crossed her arms and grimaced. "Three weeks."

Johannes and Helena shouted in unison, "That's impossible!"

"Come on, folks. You don't work on solar power because it's sexy." She uncrossed her arms and held them out wide, her spine straightening. "You've spent your whole life thinking about how photovoltaic panels could change the world. This is your chance. You need to think bigger. We need to take our wildest dreams and say them out loud." Liv turned to Hendrik. "Speaking of, Hendrik, we need a place to brainstorm. Could you book us a place to spend the next week? I want us out of the office so that we can think without distractions."

"You've got it, Liv."

"Thanks. I have to go talk to Luuk now." Liv strode to the opposite side of the floor and knocked on the door frame of Luuk's office. Her confidence from earlier began to fade. She wanted to turn around, but she could not get back to work until Luuk was on board.

"Hi, Luuk."

"What's up, Liv?"

"So, I just got off a call with Kasemchai."

"Yeah."

"He says we need to start wooing more investors."

"More investors? Why?"

She began to cross her arms across her chest, then forced them to stay at her side. *Crossed arms are a signal of defense*, she told herself. "He says that the investors will validate our business model. And that, in turn, will bring in our customers."

"All right, and increase the number of people who own our business and who we have to report to."

"I know."

"You're already hiring all these AI collaborators, and we just opened the locations in Germany and China. Why does he want to do this now?"

"It's not that he wants to start finding investors now. The first thing he wants me to focus on is developing a new business plan."

"Shit, Liv. A new one? What's wrong with the one we already presented? This is not a user experience. We're making real goddamn things!"

Liv knew everyone already had a lot to work on, but if Luuk refused, what would she tell Kasemchai? That they didn't want the funds? That she was incapable of negotiating with Luuk? She had to choose her words carefully. "Technically, our plan is very good. What he wants is to lay out a longer view of where our company could go. Something that will help investors understand the full potential our company can develop."

Luuk stared at her for a long time. Finally, he said, "He wants you to put together a vision for where our company is going—the company *I* founded. Is that right, Liv?"

"Listen, I already told him we presented a plan. This is what he just told me that he wants. I came to you immediately after hearing this from him." She raised her voice. "What do you want me to do? Be the message-taker while you two have a pissing contest over who's in charge?" She stopped and looked down at the floor. Her ears were burning. "I'm sorry, Luuk. I didn't mean to say that."

Luuk crossed his arms tightly and stared at her, holding his breath. He exhaled loudly and his face relaxed. His whole body relaxed. "That's all right, Liv. Things are moving very quickly. You're doing a great job. Let's . . . let's start over."

"Thank you." Liv slumped down in a chair. She felt both relieved and suddenly very exhausted.

Luuk jerked his head up. "Wait, why doesn't he just ask his dad for more money?"

"I don't think he has his father's blessing. If you really think about it, he always says this must be profit-driven, but I don't think he's doing this for the money. He's got all the money in the world." She slumped in the chair and stared up at the ceiling. "He's always going on about the big thing. I think it's just about doing something that no one's done before."

She looked up at Luuk. "Or maybe he's doing this to try and impress his father."

"Well, if that's the case, then we're just being swept along for the ride." Luuk sat on his desk. The fingers on his right hand drummed rhythmically. "But I suppose it's not a bad ride to draft on."

Liv grinned. "This feels like the old days when you would walk around the office and chat with people about ideas. Now it's just all work, work, work."

"That's the cost of growth, Liv. In the early days, we were getting ready for this day. It's time we finally made something with these ideas."

"Yeah, I guess so."

Luuk stared at the floor for a long moment, then, inspired, jerked his head up with a wild grin. "Liv, do you know what 'Sol-vang' means?"

"Sure, 'to catch the sun.'"

"That's right. But it's not just about solar power, Liv. The sun is the most abundant source of energy on the planet." Luuk had three quartz rocks on his desk. He picked one up. They had angular sides and a milky white center with sharp transparent tips. "Imagine this is the sun's power. Catch this, Liv," he said, tossing it to her. He picked up the second rock. "Now, catch this."

It caught the light as it flew through the air. Liv held one in each hand.

Luuk then picked up the last rock. "Are you ready?"

She laughed. “I can’t catch another, Luuk. My hands are full.”

He held his arms out, each finger stretched out wide. “Exactly. Endless energy. Free energy.” He squeezed his hands into tight balls. “But your hands are full, and you can’t handle more. That is what we need to change. The sun is not limited. We are.”

She stared out his window at the amber sky. “So, what do we do now?’

Chapter Twenty-Seven

WORKING in the evening provided some relief from the heat. When the sun rose, it was easier to see how the night wind had shifted the dust. Orange and purple bands rising on the horizon warned people to scurry into the shade.

Makoa ran down the steps of the elevated bus stop to the street. He swiped at the crusty salt that had formed on his brow before he stepped into Booch Bay, where it was dark and cool. The bar's walls were decorated with shelves that held large glass containers filled with a translucent, saffron-colored liquid. Each container was covered with white gauze and sealed with string, and a gelatinous, pancake-shaped disc floated on the top. These discs were actually living cultures of yeast and bacteria that broke down the sugars found in tea and produced a slightly fizzy, slightly sour alcohol.

When the rain stopped, people stopped drinking alcohol. It was dangerous to consume a diuretic. Water was too precious, and no one wasted a drop. But after the water began shipping from Asia and Africa, it didn't take people long to begin to imbibe again.

Producing beer and distilled spirits wasted too much water. Fermented drinks required ingredients like apples for cider or milk for kefir. There weren't cows or vineyards in the barren landscape of America. People soon discovered that kombucha, or booch, could be made with little more than water, sugar, tea, and yeast. Triple fermented to give it an extra punch. Or boosted when you wanted to laugh and forget the day.

Makoa's friends were squeezed into a booth. A glowing cylinder on the center of the table flickered like a candle. The pattern repeated itself, casting a warm glow across their faces. They laughed and drank.

Keone looked up when he approached and offered him a seat. "The waitress will be back soon."

U'ilani sat next to her brother. "Are you thirsty?" she called to Makoa. Her eyes always looked a little sad, but her face never did. When she smiled, her teeth flashed brilliantly. Her wiry hands looked strong; they *were* strong. She drove a water delivery truck. It was good exercise.

"I'm not thirsty," he replied.

"Sure, you are!" shouted Rangi. "Have a booch. Tastes better than that wine you make."

Even when Rangi shouted, his voice was soft and baritone. He always spoke slowly and deliberately. His enormous bald head sat atop his even larger body and the shiny dome reflected the lights in the room.

"How would you know what good wine tastes like?" Makoa called back.

Rangi worked in the winery supported by Philippe Labonne, the Côte de l'Est Vinery. Wine might be inconceivable for most people, but for the ultra-rich, there were the indoor vineyards.

The waitress leaned against the table; her halter top was cut low, and her shorts high. She knew all the boys by name, flirting with them and commiserating with U'ilani—the perfect friend to everyone. They ordered another round of booch, and she sashayed away.

A plate of roasted cockroaches, strung up five to a skewer, was in easy reach of everyone. The drinks went down easily, and their voices grew louder. The conversation moved from the heat to a vigorous debate on whether next month there would be more or fewer container ships of water, and as it always did in this booth, it returned to wine.

"Still experimenting with those egg-shaped concrete casks?" asked

Keone. His attention was more on the skewer of cockroaches than on Makoa.

The music was bass-heavy. Loud enough that people raised their voices and had to lean in to be heard. "Sure, we always are. Emmet just won't give up on it. He says we're always flirting with the color, and"—he put his hands up to make air quotes—"walking the line where vanillin stands between oaked and sublime."

Rangi scoffed, "Look, oak is more than flavor." He raised his glass, nearly opaque from years of use, and pointed to the bubbles rising inside. "It breathes inside the barrel. Micro-oxygenation, man. We just got a big delivery of barrels from Africa. I could sell you some for cheap."

"So what?" Makoa said. "Concrete eggs are neutral. They preserve the purity of the aromas, and"—he held his hands up and pulled them apart to stretch out the words—"the true flavors of the wine."

Makoa and Rangi had never purchased a bottle of wine in their lives; it was too precious for them. Yet, Makoa, who spent his day watching the dirt and the robots, and Rangi, who cleaned the tanks and managed the bottling machine, knew how to speak about wine. And, from time to time, they were able to taste it.

Rangi leaned over the table as best he could. "Why don't you use stainless steel or resin tanks if you want more purity?"

U'ilani slapped the table. "Always the same with you guys. Always geeking out on the wine."

"You get more mouthfeel with ceramic," Makoa pressed on. "It's more fruit-forward and more intense on the nose."

Rangi picked up a skewer of cockroaches and shook them at Makoa. "Your tannins are flat, my friend. You always want everything to taste like ripe fruit. Like a port." He glanced at the skewer in his hand, grinned and gobbled up the first cockroach on the skewer.

"Your tannins are bitter. No, astringent."

Keone slapped his forehead. "Here we go!"

"Oh, excuse me, Professor. Astringent? Besides, wouldn't you rather have a cave full of barrels than concrete eggs?" Rangi said.

U'ilani looked at the dark corners of the bar and shuddered. She turned back to the boys and said, "Sounds creepy to me."

"Well, it's about to get creepier." Makoa held up his hand to silence the group. "We have a new investor."

"Who? Don't keep me in suspense!" Rangi said.

"I didn't want to tell you guys, but I guess it's going to get out. It's Kasemchai."

Rangi's eyes grew wide. "Kasemchai? As in Kasemchai Seng?"

"The very one."

"That's awesome, man! Now we're real competitors."

"I don't know. Owned by the Asian water family? Who knows what he'll do to the winery," Makoa said. "I bet they can't even taste the difference. It's all in their head. They care more about the glass than the wine. Standing around in a wine cave drinking expensive wine like the price makes a difference."

"Like a cave full of eggs?" asked Keone.

"Shut up, asshole."

Rangi took a swig from his kombucha. "You ever notice how people like that talk about wine like it's a life-changing experience? They don't know what it's like to make wine or how much devotion goes into it. They think they can drop the big bucks." He slapped the table. "Bam, they're in the exclusive club."

U'ilani shook her head. *Tsk, tsk, tsk.* Not real words, but the meaning was clear.

In another booth, a woman shrieked, and a man laughed and yelled out something slurred and incomprehensible. In time, their voices drifted under the steady beat of the music.

Everyone at Makoa's table chuckled half-heartedly and leaned back. They stared at their kombucha, and silence slowly took over.

U'ilani leaned forward. "The candle reminds me of the beach."

Keone grunted. "It rained every day."

Each person saw their beloved island in their mind.

Makoa exhaled. "Yep."

Rangi shook his head. "They dried it all up. I never thought we would be the generation to see the island return to lava rock."

Makoa stared down into his glass. He swirled the liquid, watching the tiny bubbles excite, burst, and then fizzle. "We have no land. We have no water. What are we now?"

Rangi looked up. “It’s us against them. Where there is God, there is family. That is home.”

They all raised their glasses and tapped them together.

“Ohana!” they said in unison.

Chapter Twenty-Eight

AFTER A MONTH, Luuk, Liv, Johannes, Hendrik, and Helena arrived at Sihanoukville International Airport and were transported to Kasemchai's tower the following day. They exited the elevator on the fifty-sixth floor.

Two women with long black hair stood before them. They each wore shimmering sampots, stiff-looking silk dresses. The dresses were a captivating blend of dark blue and gold, adorned with intricate geometric patterns. The women welcomed them with a smile and a slow bow, motioning them forward with their slender arms.

As the group made their way through the lobby towards the doors at the far end, their chatter grew louder. Their excitement and nervousness were at their peak as they walked through the tall doors and entered Kasemchai's office. The crowd fell silent, taken aback by the enormity of the room and the breathtaking view of the garden and trees high above Sihanoukville. They scanned the room and took in the twisting, organic-shaped columns and sparkling chandeliers, the various seating areas, the full-sized bar with lounge chairs, and the enormous three manta rays playing over an imaginary coral mound.

Helena gasped when she realized she was standing over the glass-topped river that wound its way toward the main seating area by the great window.

Kasemchai was walking toward them over the river. “It’s an amazing view, don’t you think? Sometimes, I forget we are high above the city.”

Luuk stepped forward and shook his hand. “Yes. I don’t know what to say. The greenery. This is so exceptional, Mr. Seng.”

“Thank you. I hope your flight was smooth?” He gestured to the group.

Each person drifted forward and shook his hand.

Liv’s gaze stopped on the manta rays. After a moment, she turned to Kasemchai, a serene look on her face. “Thank you for the tickets, Kasemchai. I have to say, flying first class was an entirely different experience from the last time I flew out to this part of the world.”

“Good point, Liv. In fact, that’s where you first hooked me on this fantastic idea that we could drop solar panels onto tanker ships.” His mind also drifted back to Ishigaki Jima, and then he quickly turned his head, his face brightened, and he said, “But let’s not start with business. This evening, I have a treat for you. It’s one of my favorite sports to watch. Now, I wouldn’t be a good host if I didn’t first introduce you to a little Cambodian culture. Will you join me in the garden?”

Helena stepped forward. “You don’t have to ask me twice, Mr. Seng.”

Kasemchai winked and held out his arm. Helena hung her hand on the inside crook of his arm, and they led the group towards the outside patio.

“How do you work with this view of the ocean, Mr. Seng?” asked Johannes.

He raised his chin and laughed. “It calms my mind.”

They neared the glass wall, and Kasemchai rested the tips of his fingers on one of the twenty-foot-tall panels. It opened effortlessly, inviting them into the garden.

“It is a little warm, even at this late time of day,” Kasemchai said. “Please, enjoy the garden while I have some food brought out.”

Hendrik stared up at the giant leaf-shaped fans overhead, his mouth slightly agape.

Kasemchai watched him for a moment and grinned. “The fans

ensure there is always a breeze out here, Hendrik. And you see the mist that falls over there and there?"

Hendrik looked to where he was pointing.

"They help keep the temperature down."

Hendrik returned his gaze to Kasemchai. "You've thought of everything, Mr. Seng."

Kasemchai only smiled. Then he lifted his hand, and immediately, servants entered the garden from the opposite side and set the table in the garden's center. They were all beautiful and wore starched white shirts with stand collars and thin black suspenders. The many buttons that ran up the front were tiny and pinched the shirt out. The women wore dresses in the color of turmeric. The fabric was gathered towards the front, just below their exposed midriff.

The dishes they brought to the table were culinary artistry, as intricate as the servant's attire. A dark ceramic plate was placed on the table with something resembling a gourd, but it was soft and spongy, with delicate ribs, more yellow than orange. The top was adorned with layers of delicate fruit arranged in a fan, creating a mesmerizing flower pattern. Caramelized sugar cascaded down the sides to add a touch of sweetness. Tropical flowers and star anise floated in a liquid that shimmered with a metallic sheen. Long plates were filled with mounds of colorful chili salt and juicy slices of speckled dragon fruit, yellow-colored oranges, orange-colored plum mangoes, sourwood apples, and nearly transparent juicy round longans, each a masterpiece of presentation, a feast for the eyes.

Two more servants carried an enormous plate that was placed in the center of the table, with tiny gold stupas with delicate pointed tips. Different fish circled the outer ring, all edible. Some appeared to be made of delicate fruit carvings. Others were covered in intricate patterns or gelatinous with orange or blue scales beneath the gelatin. In the very center was a dragon that circled around its long body and stood up with an open mouth. It appeared to be covered entirely in tiny flakes of gold and wore a crown that mirrored the surrounding stupas.

Small ball-shaped desserts were served in cones made from banana leaf and dusted with coconut. Hendrik skewered one with a small bamboo fork. With his mouth still full of the sticky sweet, he asked Kasemchai what they were.

"Nom plae ai. It's a sweet rice cake with sesame seeds and palm sugar. I'm glad you enjoy them."

The group tried everything. They ate and laughed. When the dishes were cleared from the table, a tall, thin glass half filled with ice and coffee was placed in front of each person, a long silver spoon beside it. A servant poured thick condensed milk from a carafe into each glass.

She then dropped what appeared to be a pearl into the glass, and the milk chased it down and swirled like thick smoke. She stopped in front of Luuk, waved her hand over it, and lifted it gracefully toward her face. She breathed in deeply, encouraging him to do the same.

Luuk waved his hand over his glass and made a silly grin. "It smells like coffee or caramel. Or jasmine. I'm so confused."

Kasemchai laughed, lifted his glass, and encouraged everyone to take a sip.

Luuk picked up his glass, stirred its contents, and took an exploratory drink. His face lit up, and he took a second, longer sip. After one last taste, he put the glass down on the table. "Mr. Seng, before we begin our presentation today, I have to tell you, the first time Liv told me about this new business plan, I was . . ." He looked down at Liz, and a sheepish grin appeared. "To be honest, I was skeptical."

Kasemchai chuckled. So did Luuk.

"It's true. As any good business owner knows, I must be skeptical of any more dilution of my business. With you as a business partner, I felt empowered." Kasemchai gave him an approving nod. "But bringing on new investors, we have an expression, 'veel koks bederven de brij.' Or, 'too many cooks spoil the broth.'"

"Well, Luuk, I can respect your concerns," Kasemchai said. He scooted his chair back, turned his head and looked out the window. He turned back to Luuk but said nothing.

"Yes, well. That's where I was wrong, Mr. Seng. We took your advice to heart. We went to the mountain. Well, it wasn't a mountain exactly, more of a house on the coast, but you get the idea."

The group exchanged glances and laughed politely.

"We thought long and hard about, well . . ." He clenched his fist and shook it twice. On the third shake, he ended the sentence: "Power. I don't know if this is a big idea, but I can say that we will change how people think about the shipping industry."

"I'm curious. Why don't we get started then?" Kasemchai said.

Liv popped the disc off her AugWatch and placed it in the center of the table. She stood and walked over to the end. "Okay. Let's start with the original vision for Sol-vang: to reduce the fuel consumption of a tanker ship by fifteen percent with solar arrays installed on deck. We've significantly reduced the size of our solar panels and increased their efficiency, which means they take up less space." Liv flicked her hand, and a tanker ship hovered over the table. She used her hands to turn the ship and lower it so they could see the deck. "But there's only so much surface area on the deck to place our solar panels. If we could build up, then we'd have more surface area and not take up more space."

Kasemchai held his hand up in the air. He faced Liv, and then Luuk. "So your big idea is to build solar power towers?"

"Not exactly," Luuk said. "From a financial perspective, shipping companies have a compound problem. In order to reduce airborne emissions, but really driven by the rising cost of bunker fuel, they came up with an ingenious solution."

Liv turned the ship so they were looking at the stern. She raised it up so that they were below the waterline. They were staring up at the ship's enormous twin propellers. Bubbles spun off the blades and streamed past the rudders.

Luuk paused for dramatic effect. "They slowed down." The propellers visibly throttled down, changing the pattern in the bubbles. "They figured out that by slowing down a few knots, they could reduce hydrodynamic drag. And use less fuel. Their solution to reducing carbon was to run slower. The cost of fuel keeps going up, and they go slower to keep this in check."

Liv brought the ship back down, and they could see the deck again. Tall black towers descended down onto the deck. "Now, no one wants shipping to be slower. So here's our big idea: rigid sail propulsion combined with solar. Rigid sails are based on the same technology as an airplane's wing. Minute computer-driven rotation determines the optimal angle for any direction. They essentially increase lift and reduce drag." Liv pointed at the towers, which slowly rotated back and forth. "These towers will be covered with solar arrays, combining wind power to increase thrust and solar-generated energy to assist the engines."

Luuk jumped back in. "The tanker ships will be able to carry less fuel. We believe we could reduce fuel consumption by eighteen percent. Maybe even as much as twenty-three percent." He beamed at Kasemchai.

Kasemchai stared at the ship for a long time. He stood and walked around, examining the towers closely.

Luuk stood up. "And here's the best part. With the fuel savings, tanker ships can run at the speeds they were designed for. We would reduce the cost of shipping and reduce the time required for shipping."

"Do you have projections on how large the savings would be?" asked Kasemchai.

Liv motioned with her hand and brought up a chart. She positioned it in the air beside the side of the ship. "Yes, we believe this would save billions of dollars in fuel in just five years, offsetting the cost of building the rigid sail development in only four years."

Kasemchai paced back and forth, then came to a spot and slowly turned. He was searching for the words. "I'm sorry, folks. This is not a big idea."

The quiet was palpable, as if the walls themselves were leaning in to catch a whispered breath. Hendrik started to lift one of his legs as if to cross them and decided instead to freeze where he was.

"The rigid sails are unique. They're creative." He pointed at the virtual towers beside him with his open palm. "But once we've outfitted every ship, our business is just maintenance. And we still haven't solved in any meaningful way how ships use energy."

Luuk seethed. His ears were bright red. "Look, the original concept is still the big idea. All ships can deploy solar panels. Fifteen percent fuel savings is not insignificant in any way. You wanted something fantastic, and we presented a fantastic idea."

"Yes, I know Luuk," Kasemchai said. "But if shipping companies add rigid sails, we have no guarantee that they'll speed up. We'll have made a promise of returning the industry to its glory, but we can't be sure that will happen."

Luuk and Kasemchai stared at each other. No one spoke a word. The fans overhead slowly pushed the humid air back and forth.

Breaking the stalemate, Luuk began to fidget with his hands. "Look, I don't think you understand how difficult this problem is. No one has ever made a solar solution that can power an ocean-going ship. Solar panels can only generate so much power, ships are heavy and adding more weight for batteries can affect their efficiency and speed."

"So what? So it's heavy. Make it lighter. No one thought we could fly." Kasemchai was gesticulating wildly. He pointed at his AugWatch. "No one thought we could hold a computer on our wrists. Just because it's impossible today doesn't make it impossible."

Kasemchai stared hard at Luuk. He scanned the room.

Everyone lowered their eyes.

He was about to turn and walk back inside when he stopped. He looked back at Liz. "And what about your Phoenix Global Water Challenge? I thought you wanted to win that stupid challenge?"

Luuk leaned heavily on the table, his hands clasped in front of him. He lifted his right hand and slammed it down hard. "The challenge is for reducing the cost of shipping water, Kasemchai. The rigid sails are very expensive. There would be a greater decrease in fuel, but that's offset by the cost of the towers."

"Besides," added Johannes, "we don't need the Phoenix Global Water Challenge. We already have more money from your investments than we would ever get from the challenge."

"Johannes!" shouted Luuk. "That's enough."

Kasemchai stared at Luuk and then back to Liv, showing no emotion. The room was silent as he picked up his coffee and took a drink. A sly smile crossed his face. "Okay, okay. I know what you need: inspiration. I promised you a treat in Cambodia. Come with me, everyone."

Luuk glanced sideways at Liv and back to Kasemchai. If smiles could kill, his was deadly.

Chapter Twenty-Nine

THE RIVIAN BOWHEAD stopped in front of the aquadrome and lifted its great roof for the group to exit. They were quiet during the ride but began to release their tension as they passed through the doorway and into the lobby. Liv grabbed Helena when she realized the digital floor splashed water beneath each step. She watched Hendrik playfully move his toe back and forth, watching the water splash in different directions.

An attendant opened a door on the far side, and loud, deep bass music poured out.

"Come on!" shouted Kasemchai. "You haven't seen the best part yet."

The attendant led them down a long, carpeted hallway, and they emerged into a large amphitheater that was a dizzying combination of colors and sound. It was a sort of nightclub with strobing lights, bits of conversation, and laughter drifting through the air.

The ceiling was black and disappeared above them. It was dotted with vibrant rings of cloth pulled up into the center. They had delicate ribs and bands of red, orange, yellow, purple, and violet blended into each other. They undulated slowly with the slightest movement in the atmosphere, like jellyfish descending upon the party.

The group ascended shallow stairs, navigating through crowds of people who were shouting and drinking around round tables. The back of the room was lined with alcoves, and theirs was located in the very center. It featured a raised floor and was surrounded by crushed purple velvet couches. The pulsing lights accentuated their diamond-patterned stitching.

Liv stepped around the low table and sat beside Kasemchai. Finally, she could look back toward the center of the room and take in the enormous glass structure before her.

Raised ten feet off the ground was a wall of glass, easily fifteen feet high and thirty feet wide. Bubbles tumbled upwards in columns from the floor. At each end, silver baskets were bolted to the bottom with red rings encircling the top.

A man stood up from one of the adjacent alcoves and approached them. He and Kasemchai hugged and slapped each other's backs, and the men had to shout over the din.

"Annop. I think you owe me a little money," Kasemchai said.

"We'll see after tonight!" he shouted back.

The two men laughed, and Kasemchai introduced his friend to the group.

Liv leaned in and shouted over the din, "How do you know each other?"

Annop pursed his lips. "Hmm, it's complicated."

"It's simple," Kasemchai said. "We are both lovers of fast cars. Always have been."

"We are indeed, my friend," Annop said. "And after I win tonight, you'll be buying a new car for me!" He laughed heartily and returned to his booth.

A waitress approached the table, her silver hair complemented her silver bodysuit. Leaning over, she wiped her hand across the table, and a ring of drinks appeared, hovering above it. Each glass bore a unique shape, with the names of the cocktails circling the glasses.

She took their orders, and Liv realized the waitress was only gone for a moment when several waitresses returned with their cocktails. The whole world revolved around Kasemchai, and she wondered how frustrating it must be for him when something didn't go his way.

Kasemchai jumped up, stood at the front of their platform, and leaned on the tables to get everyone's attention. "You're in for a real treat tonight. Raise your hand if you've seen underwater rugby before!"

The group raised their brows at each other.

"Well, honestly, I don't know why they call it underwater rugby because there's just about nothing in common with rugby but the fighting. But you'll see for yourself soon enough. And tonight's an awesome match, FATUWR from Singapore and Club Chang from Thailand." With a Gibson in his right hand, he tapped his temple with the other. "Nothing like a good aggressive competition to get the mind moving."

He raised his glass to the group. "Repeat after me, 'Chol muy.'"

They clashed their glasses together and shouted with varying degrees of success, "Chol muy!"

He sat back down. The music was a little loud, the conversation louder. They drank and laughed until someone noticed movement behind the glass.

Helena grabbed Hendrik's shoulder and pointed excitedly, "Check it out!"

The music turned slightly hypnotic, and the room fell silent. All eyes were fixed on the woman in the water. Her golden hair flowed out from her head like seaweed drifting in the current.

She moved her legs slowly, suspended in place at the center of the giant aquarium. From the waist down, she had purple and orange scales. Where her feet should have been was an enormous billowy fin. With each movement forward, the trailing fin tips moved in the opposite direction.

The music became more tranquil, and the mermaid swam back and forth, crossing the length of the pool with strong kicks, turning in a spiral, and then stopping to stare at the crowd. Occasionally, bubbles escaped her mouth and trailed upward. Each bubble grew in size as it neared the surface.

She waved goodbye and swam up and off to the right. The music became a stronger beat, but still rhythmic and enchanting. The crowd stared at the pool, waiting for something to happen.

Five women dived into the pool, creating a burst of bubbles that rose to the surface. Like mythical sea creatures, they gracefully swirled around each other, moving in pairs and mirroring each other's movements. Their undulating bodies were mesmerizing as they twisted and spun, arching their backs in a three-dimensional dance.

They circled back together in the center. Two put their knees together and thrust their bodies down simultaneously. As they straightened up, the third woman dove down behind their legs and dolphin-kicked under and up in front of them. The two girls pushed against each other's feet and spun off each other, creating a kaleidoscope of human form. The crowd was transfixed.

Finally, the women left the pool only to plunge back in, creating a white foam on the surface that spread out in a concentric circle. Bubbles streamed off their lithe forms as they performed their finale. The room was silent, and then erupted with clapping and shouting.

The music stopped, and the announcer broke in. He started slow, knowing the crowd had been lulled into a trance. In minutes, he had turned the mood around, finally reaching a crescendo as they cheered for the main event.

This evening's event included two matches, each with two fifteen-minute halves. The referee, with a micro scuba tank in his mouth, placed a rubber ball on the halfway line at the bottom of the tank. It was the size of a water polo ball but filled with salt water to make it denser than the surrounding freshwater.

The buzzer sounded, and players crashed into the water. The monitors to the left and the right showed the teams from each side furiously swimming towards the ball. Their black fins pumped up and down like dolphins. Bubbles streamed behind them. The men wore tight briefs. The women wore one-piece or two-piece thong swimsuits. On their heads, they wore rugby caps, scuba masks, and short J-type snorkels.

One of the players reached the ball first. He curled over it and immediately threw it to a teammate. Instantly, the players were shooting in every direction, swimming around and around, tugging at the ball, knees bashing into their opponents.

A woman intercepted the ball and shot toward the goal. She swam over another player, then under the next. Just as someone grabbed her shoulder, she tossed it to her teammate, who shoved the goalie out of the way and slammed the ball into the basket.

The buzzer went off. The crowd leaped to their feet and cheered.

Chapter Thirty

THEIR CAR NAVIGATED WET STREETS, finally stopping in front of a narrow alley it couldn't pass through. Halfway down the alley, a row of red lanterns with delicate ribs beckoned. They were somehow both beautiful and gaudy.

Kasemchai led the way. “Does everyone like Japanese food?”

“I love it,” Liv said. “So fresh and healthy.”

“Well, this is the best Japanese food you'll ever have. My mentor and I came here often. Delicious? Yes. Healthy, well . . .” He grinned as he pulled back the curtains that covered the door and motioned for them to enter.

Inside, chaos surrounded them. People yelled and smashed their beer glasses together. The waitstaff turned towards the door and shouted in unison, "Irasshaimase!"

A young man wearing a white T-shirt and a white towel wrapped around his head shouted hello and rushed over. It seemed he knew Kasemchai well, as he made jokes about his clothes and playfully pushed him to a table. The rest of the group followed and had just begun to settle in when he tossed menus on the table and walked away.

He returned with three large beers and popped the bottle caps off with a pair of chopsticks, shooting the caps across the room to bounce off the walls. One hit a man in the back of the head. He turned, grinned, and raised his glass toward the group.

Kasemchai grabbed all the menus and shouted over the din, "Don't worry about ordering! I will order lots of food. The idea in an izakaya is to eat and drink, drink and eat."

He turned to the server, and the two huddled over the menu. They pointed, debated, laughed, and finally, the server looked pleased and returned to the kitchen with their order.

Another server poured everyone's beer and took away the empty bottles. She returned a moment later with two more bottles and immediately opened them, shooting the bottle caps across the room.

Kasemchai turned to Hendrik and shouted, "What do you think of the beer? It's not fancy, but it goes down well."

"I love it," Hendrik said. "I've never had it before."

Kasemchai cocked his head, and gave him a crooked smile. "What do you mean, you've never had it?"

"We only drink booch. I've heard of beer, but it hasn't been available in the Netherlands for a long time."

Kasemchai laughed. "Then you are going to have a wonderful time tonight. You may hate me tomorrow, but tonight . . ."

The two men laughed and crashed their glasses together.

Another server placed several ceramic bowls filled with green edamame. They sucked on the salty soybean pods and popped the beans in their mouth, driving up their thirst for more beer. A couple of plates came with small chunks of sizzling meat or vegetables cooked over charcoal on short six-inch skewers called yakitori—a half-dozen parts of the chicken, including crispy skin and meat from the neck. Chewy ginkgo nuts, pork, sausage, octopus, and green onions, all lightly seasoned with a sprinkle of salt and a squeeze of lime.

Just when they thought they couldn't eat anymore, they served dishes of fried chicken karaage and okonomiyaki, a savory pancake slathered in a sweet and salty sauce and mayonnaise drizzled in lines across the top.

Kasemchai took a deep drink of his beer, slammed it down on the table, and loudly asked Liv across the table, "Why don't you just make bigger solar panels?"

She cocked her head to the right and looked off to the left. "We could make slightly bigger panels, but creating energy isn't the challenge. It's storing it. We can optimize our solar panels to thirty-two percent efficiency, but that number drops immediately with clouds and as the sun sets."

"Well, why don't you make higher-density versions of those saltwater batteries on ships."

Liv laughed. "You mean sodium-ion rechargeable batteries."

"Yes, those. Why don't we just fill the ship with those things?"

"Well . . ." She placed her hands on the table, one on top of the other. With a look that implied she had explained this to him numerous times, she said, "Because our model is about using solar to reduce energy consumption."

Kasemchai leaned across the table. His eyes widened, and with an emphatic voice, he said, "Come on, Liv. Go bigger."

"Okay, let me think."

"You do that," he said, and grabbed a chicken skewer.

"The ships already have electric engines. What if we made batteries so powerful that you replace the entire engine?"

"That's exactly what I'm thinking," Kasemchai said. He threw his arm out wildly when he said, "Just rip out the entire gas engine."

"But if we're going to replace this bunker fuel, we must develop a lighter, safer, easier-to-manage, and cheaper solution."

"What if it's just lighter?" asked Kasemchai.

Johannes leaned in. "Not good enough. Ships must refuel whenever they're in port, but the complicated infrastructure already exists. We have to be better in every way. Or no one will change."

Liv raised her hand. "Wait, wait, wait. What if we completely replaced a ship's engine with a solid-state battery? A ship could make two trips without charging in port. When the ships are in port, we recharge them. There would be a big upfront cost, but then it would be cheaper to run for the remaining life of a ship."

Luuk placed his elbow on the table and leaned heavily over it. "Which could be twenty or thirty years. It's technically possible. But marine batteries have to operate in harsh conditions—extremely high and low temperatures. And the salty air is very corrosive. Combustion engines have to be constantly maintained . . ." He sat up straight and rested both hands on the table. "But the shipping companies are used to this kind of maintenance."

"Got it," Kasemchai said. "It's a race to show ship owners that they can't overlook the cost of our solution."

"That's right," Luuk said.

"So, how do we win this war?" asked Kasemchai.

"It's funny you should say that, Mr. Seng," Johannes said. "I have a second idea. Military vehicles sometimes run on gas, but mostly diesel. As you might guess, very powerful combustion engines are reliable, but they're very hot, noisy, and spew black exhaust. And an internal combustion engine means, by definition, that it needs to run on something explosive. They work very hard to protect the fuel tank. No one wants to be in a burning tank."

He tugged on a grilled octopus tentacle and chewed vigorously before he continued. "And they have another problem. When an army is mobile, they have to store all that fuel in—get this—giant bags. Imagine how hard it is to move enough liquid to support all those vehicles."

Luuk glared at Johannes. "Are you suggesting we work on military projects now?"

"If we are going to change the world, we need to help everyone who uses dirty fuel," Johannes replied. "I'd like to see them end their reliance on fuel. Imagine setting up mobile solar systems that could recharge batteries anywhere in the world."

Liv jumped in. "We catch the sun and store it in solid-state batteries that could safely be used in tanker ships and military vehicles. We build mobile solar grid stations that can be made in any port or that the military could install temporarily or permanently."

Luuk looked around at the smiling faces. "I hate to be the bearer of bad news. These are all great ideas. But you're overlooking the immense cost of overhauling all of these ships. We're talking about the harshest environments. In order to make them reliable, we need to monitor them closely, and the military would never allow that."

Everyone slumped in their seats. Johannes sipped his beer, and Liv poked at the eggplant in front of her. Kasemchai lifted the glass and drank the rest. He stared into the bottom as the bubbles dripped down toward his mouth.

Suddenly, he slammed the glass down on the table. "Of course!" he shouted. "The solution is 'BaaS.'"

Luuk glanced down at the grilled fish. "Bass? Like the fish?"

"No. B-A-A-S. As in battery-as-a-service. We don't sell them the batteries. They sign up for an annual service. When the ships are in port, we remove the batteries with the same cranes used for loading containers and drop in fresh batteries. That way, we can keep them clean from the salt air."

Johannes jumped in. "And the batteries don't need to be monitored, which makes the military happy. We just swap battery packs and run diagnostics on the batteries to ensure they're working perfectly. Charge them up, and they're ready for the next trip."

"Won't this require a lot of infrastructure?" asked Liv.

"Sure. But we'll create service stations in every major port. Maybe we'll license the technology to partners in smaller ports. When ships come in, we'll swap the batteries with new batteries. And Luuk—" Kasemchai stopped and smiled.

"Yes?"

"This all runs on renewable energy. Whatever makes sense. Solar energy in Shanghai, wind energy in Oakland." He ran his fingers through his hair as his mind raced. "Everyone is risk-averse. We take the risk out of the equation by keeping the batteries healthy. And we have a business interest in improving the charging capacity and strength of our batteries. The first company to create the infrastructure that charges and loads batteries in every port would be the biggest shipping partner. This is as important as water. We are ready, Luuk."

Luuk looked around the table like they had all gone mad. "We're ready for what?"

"The Phoenix Global Water Challenge!" Kasemchai shouted. "We're going to submit our idea for the challenge. This will be the solution that ships more water than any idea ever pitched before."

Luuk's face turned purple, and a vein clearly stood out on his neck. "Do you even know how technology works? This could take years to develop. And we don't even know if it will work."

Kasemchai wasn't offended. He didn't protest. He simply grinned before saying, "Luuk, I've never met a problem money couldn't solve. You just have to decide that you want change to happen." Kasemchai waved at the server, who returned with three new brown bottles of beer.

Pop. Pop. Pop. The server topped off everyone's beer and then poured a glass for himself. He raised his glass into the air and, with a gravelly voice, shouted, "Kampai!"

The whole table shouted, "Kampai!" and clashed their glasses together.

Chapter Thirty-One

PHILIPPE STOOD on the path leading into his garden. His long lilac-colored jacket contrasted the many shades of green behind him. His eyes were closed, and he looked serene as a woman beside him carefully dabbed at his face with a large sponge. She hummed as she carefully removed the last few shiny spots.

Angélique sat at a table with the director, studying a virtual bank of monitors. Two of the monitors showed Angélique's father from different angles. A third was pointed up towards the ceiling, capturing the birds as they dropped from the ceiling and then swung back up as if they had reached the end of an invisible cord.

The script floated to the side. She nudged the script over and pointed to the second paragraph. "When he says '. . . and where does the rain come from?' I want the camera drone angle to follow his arm, up past his finger, and fly up towards the sky."

The director excitedly added, "Like we're following his gaze up to his line of sight. Got it?"

"Right! Pick a light and zoom up to all-white. Then we'll cut to his face."

The director looked up into a space in front of him. "Okay." He made a few swipes with his finger, spoke to his companion, and seemed pleased with the edits made to the script.

"That's right," Angélique responded. "I want the focus on his face. This is the key message. Slowly back up to fill half the frame and walk with him."

The director scrolled down the page. "Okay, and here's where we cut to his side. He bends over to pick up the glass of water."

"Right. He picks up the glass and goes on about the taste. Slowly pivot around him to end on his half frame. 'Delicious water, the kind you can only taste when it comes from above.'"

"Okay, Angélique. I think we're ready to go."

Chapter Thirty-Two

After work on Thursday, Makoa headed to Booch Bay. When he entered the bar, Rangi, Keone, and U'ilani were already seated in a booth. U'ilani scooted over to make room for him when she saw him approach. He returned her smile with a dour grimace and slid into the booth.

"Are you thirsty?" asked U'ilani.

He laid his hands flat on the table and sighed. "I'm not thirsty."

"Wow, that was depressing," Rangi said.

The waitress approached them, leaned over, and planted her hands firmly on the table. She smiled at Makoa. It was one of those smiles that made him feel special, like she had a crush on him. He knew it was just business, but he couldn't resist a sly wolf grin in return.

"What can I get you, Makoa?" she asked.

Rangi held up his hand. "I'll have another—"

She slapped his hand down playfully. "You wait your turn, Rangi. The poor boy just came in."

Keone burst out laughing.

"Thanks," Makoa said. He gave Rangi a playful push. "I'll have the ginger booch."

She looked back at Rangi, grinned, and said, "U'ilani, what would you like?"

"Hey," protested Rangi.

"Oh, don't you worry about a thing, sweetheart. You know you're always on my mind."

Keone and U'ilani ordered their kombucha, and finally, it was Rangi's turn.

"Do you have any watermelon booch?"

"I always keep a stash for you, Rangi."

"Is it made with real watermelon?" he asked.

Her eyes grew wide, and then she leaned in towards Rangi. "Well, honey, our drinks are as real as the smile on my face. I have to admit, the watermelon in them is more synthetic than your virtual girlfriend. But trust me, just as fun and sweet!"

The table erupted in laughter. She sat down in the booth next to Rangi and rested her head on his shoulder. "You know I tease you because I love you. Now, can I get you a watermelon booch?"

"Yes, ma'am."

"And another boost?"

"Of course. How else am I to put up with your flirting?"

She slapped him on the chest and protested, "Rangi!" Then she jumped up and headed off to put in their order.

Keone regained his composure and looked over at Makoa. "How's it going with your new investor friend?"

Makoa scowled. "Do you know he flies out here regularly just to visit the winery?"

"And poor old you don't even have a car."

Makoa punched Rangi in the shoulder, who drew back in mock surprise, then rolled his eyes and drank his booch as if nothing had happened. "But seriously, he's got all the money in the world. He doesn't appreciate the wine."

"Well, he drinks the wine," U'ilani said. "Apparently enough to become part-owner."

"Sure, he drinks wine. But he keeps talking about how sexy the wine is. And how it needs to be sexier. Wine should be appreciated just for what it is."

Rangi stole a glance at the waitress, a dumb grin on his face.

"And what is that, little brother?" asked Keone.

"Wine is nectar from the gods," Rangi said.

"But," Makoa said, "it should be affordable. It should be something that anyone can drink. I know it's too expensive now, but we were going in that direction. Now, he just wants to make these ultra-reserve wines."

"What the hell is an ultra-reserve?" asked Rangi.

Makoa slammed his fist on the table. "Absolutely nothing! There's wine, and there's the vintner's reserve wine. He keeps saying 'ultra' is sexy. Everything is sexy to him."

"What's this really about, little brother?" asked U'ilani. "You can't just be upset that he calls all the wine sexy."

Makoa wrapped his hands around his cup and stared at it, lost in thought. At last, he leaned back and turned to Rangi. "I don't know. How did you get Philippe to invest in the Côte de l'Est Vinery?"

Keone laughed out loud. "You think Rangi had something to do with that?"

Rangi's eyes bugged out. "Hey, man. I'm an important person. Those bottles don't just drive themselves to the aging caves."

"It doesn't matter. What matters is that Philippe is a man of God. And that's what's important."

Rangi deftly pointed to his head, his stomach, his left shoulder, and then his right. "Amen to that."

"We all do the grunt work," Keone said. "What's important is that we know that our leaders care about the one God."

"Amen," U'ilani said. "I hear what you're saying, Makoa. It's a shit world. We don't need someone from another country telling us what to do."

"What he needs is a reason." Makoa paused, and everyone turned to him, bobbing their heads imperceptibly to the heavy beat of the music. "A reason to listen to the people."

"No," Rangi said. "What he needs is a lesson. We're just pawns in Kasemchai's grand game. People like that decide the direction of the business, and we're here to carry out their plans."

"We're just cogs in their wine-producing machine!" Keone shouted. "We need to fight fire with fire. Clean out the bad."

"That's right. You can't appreciate good wine until you've had bad wine." Rangi leaned in and grinned. "And we've had bad wine,"

U'ilani raised her glass. "To the bad wine!"

The others raised their glasses and shouted, "To the bad wine!"

MAKOA LEFT the bar earlier than the rest, but the sharp pain in his temple was reminiscent of an all-nighter. Rubbing his temple, he walked slowly from the elevated bus stop to the winery, staring at his feet.

His nose twitched, and he could smell the acrid smoke before he saw the building. The intumescent coating was charred, making the building look puffy. The air was still, and smoke drifted straight up and blended into the sky.

Makoa's heart raced as he sprinted toward the building. Maybe there was something he could do, but he was already coming to the realization that the flaking sides of the building and the smoke meant the fire was gone. He stopped when he saw Emmet standing alone, staring up at the building with his hands on his head.

Makoa approached him slowly. He wanted to ask what had happened, to share in the shock of this tragedy, but he had a feeling he already knew the cause of the fire. His stomach churned in a knot that wound tighter and tighter.

"Hey, Emmet . . ." Makoa placed his hand on Emmet's shoulder for a moment, then stuck his hands deep into his pockets. "What happened? Was there anyone inside?"

Emmet's slow response was unusual, almost like he was under sedation. "No one was inside at the time. The fire started last night in the back of the building."

"Do you know how it started?"

"I don't know." Emmet sighed heavily and kicked at the ground. "Someone moved a trash bin up against the building and lit it on fire."

"Why would they do that?"

"I don't know. There were three people." Emmet turned and looked at Makoa directly in the eyes, like he knew it was Makoa's friends. Then he turned and faced the building again. "It's all on camera, but they wore masks."

"I'm sure the police will figure it out, Emmet. There must be a clue."

"I don't know . . ." His voice trailed off.

"I can't believe it," Makoa said. He wanted to say something reassuring but didn't know how to comfort him. He recalled the ache as his jet lifted from the ground in Honolulu in a direction that could never be reversed. The realization that a shell of him was being left behind.

Emmet stared at the ground. He didn't answer Makoa.

A tiny, desperate part of his brain thought he could unwind the day. "Have you spoken to Kasemchai yet? Will he rebuild?"

Emmet slowly turned to look up at Makoa. "Rebuild?"

"Yeah, he owns half the company, doesn't he?"

"He's an investor in the company, Makoa. The winery—" He swept his arm out towards the building. "There's no more winery. There's nothing to invest in, Makoa."

Makoa stared at him for a long time. Then he realized he was staring and looked down at the ground.

"I'm sorry," Emmet said. "I'll make sure you all get paid through today. I'll see if I can come up with more money for the rest of the week."

"We'll all pitch in," Makoa said. He started to walk toward the building. "I'm sure we can all help clean things up."

Emmet grabbed his arm. "You can't clean this up. There's nothing left. I think . . ."

He dropped his hand. Makoa was certain he would say something more, but he simply turned and walked away from the building. It was like he was going somewhere, and then he forgot where he was going.

Makoa thought about his little girls. He tried to imagine what he'd tell his wife. There was nothing to say.

He knew his friends had set the fire, and he knew why they did it: to teach Kasemchai a lesson. When he left the bar, they were laughing and saying stupid things. This was entirely their idea.

But he hadn't told them to stop talking nonsense. He might as well have lit the fire himself.

He wondered why everything always ended in ashes. He placed his palms on his eyes and pressed until they hurt. There was nothing else to do.

He turned around slowly and walked back to the elevated bus stop.

Chapter Thirty-Three

MORE THAN A YEAR had passed since the meeting in Sihanoukville, and Kasemchai was landing in Rotterdam. The previous year, Liv had moved Sol-vang into the much larger space in the World Port Centre, a tall building on the tip of the same peninsula. It was rounded on the west side and had two floors on top for special events, making it look like an enormous submarine conning tower.

He rode up the elevator of the World Port Centre and smiled immediately upon seeing Liv as the doors opened. "Good morning, Liv. Are you thirsty?"

"No, I'm not thirsty. What do you think of our new location?" she asked as she waved her arm in a sweeping motion.

"It's fantastic. The old office was so fun, but you needed the space. But you know what I don't love? The heat."

"I know," Liv said, looking out in the distance. Her smile dropped momentarily, and then she caught herself. "I've seen pictures of my mother playing in the snow. It looked magical."

Hendrik leaned over the railing above and waved to Kasemchai. "Good morning, Kasemchai. So good to see you again."

As Hendrik descended the stairs, Kasemchai turned back to Liv. "You guys have really moved up since I saw you last in Sihanoukville."

Liv laughed. "I know. It takes me longer to get to this floor than it does to get into downtown."

Hendrik was a tiny bit winded when he finally reached Kasemchai. "Follow me. I'll take you to our conference room."

The hall was painted floor to ceiling with a scene of a canal when there used to be water, giving a sense of cruising by houses above and boats moored to the side. When they entered the room, Liv pointed to one of the tall-backed chairs for Kasemchai to sit in. She walked to the front of the room and remained standing.

"Would you like some tea, Mr. Seng?" asked Hendrik.

"Do you have any black tea with a little sugar?"

"Of course. I'll get it for you. And Liv?"

"No, thanks, Hendrik." She waved her hand over the light panel on the wall to darken the room. "A glass of water, please."

"Sure thing." Hendrik left the room.

Kasemchai looked over at Liv. "So, how are things?"

Liv tapped her AugWatch and scrolled through the colorful bubbles, pointing and swiping. The virtumersion activated, and the word "Overview" hung over the wall in white on a black background. "Well, there are two areas we're testing—"

"I meant with you," Kasemchai interrupted her. "How are you doing?"

"Oh." Liv's ears glowed. "I guess it's going well. Busy, as you can imagine. I play racquetball now."

"Really, in this heat? Doesn't that make you sweat?"

Liv looked up and then down quickly again. She smiled at her shoes. "It's indoors. I always wanted to learn but never had the time. Well, I guess I don't have the time now either. But I needed to find a distraction." She gave him a quick smile. "I'm quite good, you know. Do you play?"

"A little. We'll have to find some time to play while I'm here."

"Is there anything the fantastic Kasemchai can't do?" asked Liv.

Kasemchai chuckled and stared directly at her. "We shall see. I don't often play these days, so you might rub me all over the court."

"Is that a thing?" asked Liv.

"No, no, no. I may have said that wrong."

They both laughed.

At that moment, Hendrik entered with their drinks. He walked over and placed a cup of tea in front of Kasemchai. The heavy white ceramic made a *clunk* sound as it hit the wood. A moment later, he left and closed the door behind him.

Liv swiped up on the light in front of her AugWatch and began her presentation. "The development phase is going well. The technology has been around for a long time, but no one has come up with a practical use for it. The focus has always been on lighter packs for cars and planes. You can see here—"

"Sorry, are you showing something?" Kasemchai interrupted her.

Liv glanced down at her presentation. "You don't see my intro?"

"No."

"Noa, why isn't my presentation showing to everyone?"

Noa's disembodied voice said, "I'm sorry, Liv. I didn't realize you wanted to show your presentation to everyone."

Liv pointed to the device on the table. "But it's on the experience sphere."

"Yes, but you said you were testing. I thought you were testing your presentation and didn't want to share it with the group. I'll show your presentation to everyone now."

Liv looked up at Kasemchai. "Now then, can you see the word 'Overview'?"

Kasemchai gave her the thumbs up.

"Okay, as I said before, we're using an ultra-dense solid-state battery. We've designed a more durable pack sealed in a material that can withstand continuous exposure to a ship's saltwater environment." She swiped her fingers to show a rendering of the battery. It highlighted three parts: the anode on the top, a titanium separator, and the cathode on the bottom. Arrows flashed to show the ions flowing from the anode to the cathode and electrons flowing from the cathode to the anode. She used her finger to point an arrow at the anode side. "Well, the titanium-based anode battery is working very well. With over three thousand charges."

She looked over at Kasemchai. "Do you know how we made the

sodium-ion battery work? All those engineers we hired, and it only took one. She was able to talk to the AI about the future. She imagined the future and didn't stop asking questions until it made it possible." She gazed off into the distance, making calculations in her brain. "I thought the anode would be silicon, but we couldn't solve the volume expansion issue. Titanium. Amazing."

She flushed and returned her focus to Kasemchai, realizing she had gone off into her own world. "I'm sorry. Anyway, it's still bigger and heavier than our goal. This makes it hard for military applications at this time, but we can wire enough in parallel to power a Panamax-sized tanker."

"I appreciate the engineering challenges, but that's not big enough," Kasemchai said.

"I know. Our target is the Very Large Water Carrier, the VLWC. That's a jump from a hundred twenty thousand deadweight tons to over three hundred thousand deadweight tons. We can link more batteries than we currently designed for the Panamax size, but we need to increase their energy density. This will take time. As I was saying . . ." She swiped her fingers to show an illustration of a large, flat building. "We explored contracts with several gigafactories in Germany and ultimately decided on the Northvolt Drei gigafactory in Hamburg."

She walked Kasemchai through a virtual tour of the factory. Then she zoomed out so that he could see a map of the location. "We expect to make three hundred thousand tons of cathode material and a hundred thousand tons of anode material annually—here."

Kasemchai stared at the image of Hamburg. He rubbed his chin. "Hmm. And why Germany?"

"There are already quite a few battery factories in Germany. There are skilled workers and plenty of land."

"Uh-huh. And where are most of the ships made these days?"

"It doesn't really matter where they're made because we'll be retrofitting ships. And we get a lot of traffic on this coast."

Kasemchai took a sip of his tea. "Sure, Liv. But think bigger. Eventually, all ships will be made with batteries instead of engines. So, where are most ships made?"

Liv grabbed the back of her head with her hand and held it there as if trying to hold it from flying backward. "Do you know how much work we've put into finding a European location? We've already started extensive negotiations with the government."

"I know, Liv. But ships are made in Asia. The biggest locations are in Japan, Korea, and China."

"Yes, but ships move. We'll focus initially on ships that can stop in the ports of Amsterdam, Antwerp, Hamburg, or Rotterdam. We can ship batteries from there to any other port." She realized she had stepped out into the sun and stepped back into the shade under the restaurant awning. "But the main reason we're building them here is it's close to several other battery factories, so we have a large base of robots and qualified AI collaborators to work on the project. No location in China can turn on production for these specs right now. There are already established distribution centers for materials, and there are rail routes directly to the ports where we can retrofit ships. They're eager to open them up for new construction, as long as it's clean construction."

Kasemchai nodded as if he understood. His stomach churned. He knew that Liv had thought of everything, but he also knew that Mr. Sato would want this operation to be in Asia. "You seem to have thought it all out, Liv. As you always do."

"I'll do more research, Kasemchai. See if I'm overlooking something."

"Please. And keep me up to date on this part. I want to be absolutely certain before you kick off a major project."

THE NEXT MORNING, Johannes barged into Liv's office. "You're not going to believe this. I have a meeting with the Bundeswehr. They're considering a pilot of our batteries for a German military vehicle."

Liv leaned back in her chair, a look of surprise on her face. "Already?"

Johannes dropped his hands and shrugged. "Already? That was like a year's worth of work to get in the door."

Liv pushed her chair back and stood. "No, no. That was not easy. I'm just surprised that our first contract would be military."

"Well, we don't have a contract yet."

"What's the pilot for?"

Johannes looked down at his AugWatch and asked his companion to show Liv what a Fuchs looked like. The vehicle appeared in the room with three rugged tires on each side that were nearly up to Liv's shoulders. It was painted with camo and had a sharp angular design. The only real window was on the front and the hood sloped down in a flat wedge that looked like it could easily drive through anything it encountered.

"Their armored personnel carrier, the Fuchs."

Liv approached the enormous virtual model cautiously. "I'm glad to know it's for a people carrier, not a weapon. Is it for the power pack?"

"No, complete engine replacement."

He pointed to the front right of the vehicle, and the engine appeared behind the now transparent armored wall. He waved it away, and it was replaced by a soldier sitting in a seat.

The height of the vehicle rose, and a long rectangular battery appeared along the floor of the vehicle. "We had to raise the height of the vehicle six inches, to place a battery on the chassis. This lowers the vehicle's center of gravity, making it more stable."

"Wow. This is a good test. And they'll pay us?"

Johannes laughed. "Yes, Liv. It would be a paid pilot. I think a couple of things put us over the top."

"And they are?"

"We said we could reach almost the same range as their diesel engine. Battery propulsion would make them quieter, which reduces the crew's fatigue and makes them stealthier."

"What about our battery composition?"

"Of course. The Fuchs are amphibious, and we have a water-friendly battery. And most importantly, our commitment to building the batteries in Germany."

Liv turned her head quickly, and her eyes narrowed. "We haven't closed on the contract with Northvolt Drei gigafactory yet. And I'm getting a lot of pressure from Kasemchai to build batteries in Asia."

"I know," Johannes said. "But now you have another good reason to start in Germany—government subsidies."

Liv sighed and looked out the window at the night sky. "Okay, I guess it's time to start closing deals, Johannes. You got yours. I'll get mine."

She started toward her office door.

Johannes stood where he was. "Liv," he called to her.

Liv stopped and turned back. "What?"

He started to speak, and then glanced down at his feet. Finally, he looked back up at her. "Thank you, Liv."

She tilted her head to the right. "Of course."

"No, I mean, thanks for keeping me. I . . ." He wrung his hands together for a moment before continuing. "I know I was always kind of a prick to you. Thanks for keeping me on the team."

Liv smiled broadly. "You were a prick, Johannes. But you also helped me be better. And you have good ideas. That's why you're on my team."

Johannes shoved his hands deep in his pockets and grinned back.

"Come on. Let's go break the news to Luuk."

LUUK'S thunderous voice could be heard through the door and down the hall. A dozen hands cautiously dialed back the immersion setting on their AugWatch to see who was in Luuk's office.

"Crap! Our first official contract is with the Bundeswehr? To help German soldiers fight? We're trying to save the Earth, Liv. 'Sol-vang!' To catch the goddamn sun, Liv. I chose that name to help people understand how we could save the Earth, not blow it up."

Johannes stepped forward and said, confidently, "It's still the same thing, Luuk. Military vehicles use a lot of fuel and create tremendous pollution."

Luuk jerked his head toward Johannes and stared him down. When he spoke, his voice was quiet—horribly quiet. "Did I ask you a question, Johannes?"

"I . . ." Johannes spun his head to look at Liv and back to Luuk. "No, sir."

"Then keep your goddamn mouth shut." Luuk turned to face the window, ran his hand through his hair, and continued turning until he was facing Liv again. "I'm disappointed, Liv. I thought we had an understanding."

"An understanding about what?" demanded Liv.

"You're the president, and I'm the CEO. I'm still driving the vision. But what am I hearing? You have meetings all the time with Kasemchai to discuss strategy without me. Johannes is off talking to the German government to help wage war. Who knows what the hell else you're working on?"

"That's not fair, Luuk." It wasn't a complaint. Liv knew how to be tactful, but she was not going to back down. "I've kept you informed of all activities. And, of course, we've been talking to the German government, we're working to open our battery factory there. There's not just one team to work with. We're following many paths."

"Many paths? You tell me things, but I don't know what you're doing."

"Luuk, this is a paid contract. Johannes has been working on this for over a year," Liv said. "And it's going to pave the way to finalize our contract with the Northvolt Drei gigafactory. I came here to let you know in advance so I can close the deal."

Luuk walked up to a shelf on his wall. He had several crystals on it, like the ones he'd shown Liv before the trip to Sihanoukville, but larger. There was an amethyst geode with its sharp edges pointed toward the center and bands of purple that were darkest in the center. And a crystal with sharp repeating geometric shapes and brilliant rainbow colors.

He lifted the amethyst geode and examined it. Liv and Johannes' eyes darted back and forth at each other, assessing the uncomfortable silence in the room. Luuk slowly placed the crystal back on the shelf. He didn't turn around.

"You don't need my permission to close the deal with Northvolt Drei. You're the president, Liv." He looked up at the ceiling and back to the shelf of crystals. "Anything else?"

"That's it, Luuk. I'll keep you up to date on our progress."

Liv and Johanne shifted on their feet uncomfortably until Luuk quietly said, "Thank you."

THAT WEEKEND, Liv and Brechtje stood on the tip of the cliff overlooking the Sputnik Crack in Switzerland. The grass was tamped down from previous base jumpers in patches between small yellow flowers.

The two women looked at each other, briefly grabbed hands, and counted down: "Three, two, one!"

They immediately picked up speed, nearly a hundred miles per hour in less than ten seconds. Head-first, their bodies were parallel to the straight cliff wall until their wing suits generated enough lift to slowly move them out over the lush green grasses and trees of the Alps.

The wind howled in their ears. They seemed to be slowly hovering until they passed closer to the tops of the trees and could feel the ground flying under them. Touching a tree or the tip of a rock would shred their bodies in seconds.

The ground leveled out as they approached the Sputnik Crack, a gouge in the side of the mountain that was their shoot down toward Lake Walen. Liv was mesmerized, staring at the crack, and then she realized she was dropping down into it.

She moved ahead of Brechtje, gave a triumphant yell, and flew into the gorge. It was ridiculously narrow, and the updrafts pushed her from side to side. Just as she tried to pull back from the wall on the left side, Brechtje slammed into her, and the two girls screamed and tumbled over and over.

Liv bounced on the ground, then lifted for just a moment before rolling onto her back and falling again. Brechtje's helmet crashed into hers, and the two girls howled with laughter.

"Dial back the immersion setting, Noa," Liv said. She looked around at the glass walls of the vertical wind tunnel.

Brechtje held her arms and legs out like a starfish and rose back up towards the ceiling. Liv jumped up, caught the air, and shot towards her. The vertical wind tunnel turned off, and the two women bounced along the floor toward the door. Their legs shook as they wobbled toward a railing.

Liv removed her helmet and wiped the sweat from her brow. "My god, that was intense. I nearly had a heart attack when you slammed me into the wall."

Brechtje hung her arm over Liv's shoulder. "You're getting better. We almost made it out of the gorge that time." Brechtje pulled out a terry towel from her gym bag and wiped her face and neck. Out of an old habit, she started to squeeze the sweat from the towel into her mouth and caught herself. She laughed. "Can you believe I was . . ."

"I know exactly what you were going to do. Let's hope we never see days like those again."

"Agreed. Time for a drink?"

"Yeah," Liv panted out. "I need to rehydrate."

The women walked downstairs to the café and ordered iced mint teas. Liv dropped down in her chair, exhausted. She lifted the glass to her face and inhaled deeply. The fresh mint leaves submerged in the tea were intoxicating.

"Brechtje?" she asked.

"Yeah?"

"I need to make someone happy. But I can't give them what they want."

"Okay." Brechtje leaned in close, a mischievous grin across her face. "What does he want?"

"It's Kasemchai."

"I figured. Just tell him, yes, you'd love to live with him."

Liv laughed. "No, it's not like that. I'm working on a project in Europe, where it should be, but he wants it to be based in Asia."

"He trusts your judgment." She sat back in her chair, crossed her legs and held her hand out like she was presiding over her court. "And he knows you've done the research, right?"

"Yeah. He has a good reason, but he doesn't understand the process of what I'm building. And there are other dependencies."

"Okay, my cryptic friend who won't tell me the details of what you're working on. Then maybe you *should* do it his way."

"From the starting point, it's wrong. But in the final stage, he's right." Liv stared intently at her hand and drummed her fingers on the glass table. "Maybe that's it. I can still make it where I think it's right. But then we'll do the last phase in Asia." She stopped and looked up. "Brechtje, you're a genius!"

"Well, of course I am. Do I get to charge you a consulting fee now?"

"I think you should. But you have to pay me back for slamming me into the Sputnik Crack."

Chapter Thirty-Four

KASEMCHAI ENTERED his office and immediately noticed the large box wrapped in red paper on his desk. He unwrapped the paper to see a red rose-colored box inside. Carefully, he lifted the lid to reveal a tea set: three small cups and a small teapot. The Celedon ceramic was a milky green-blue with delicate crackling under the glossy glaze.

He picked up a folded card and opened it. Handwritten inside was the simple message: *Call me. Liv.*

Kasemchai smiled to himself. "Hey, Chanthavy, call Liv," he instructed his AugWatch.

Liv's avatar appeared in the room. It walked over to the chair across from him and sat down. "Hello, Kasemchai. Did you get a gift?"

"Yes, I did. Care to tell me why I have a tea set?"

"Okay, hear me out."

His smile dropped from his face. He began to fear the gift was an attempt to sugarcoat news he didn't want to hear.

"You asked me to see if we could move our operation to Asia. My research confirmed there's really no better location to build our batteries than the location I originally indicated in Germany, but . . ."

"Is this a good but?"

Liv's avatar looked up, nonplussed. "It's a great but."

Kasemchai grinned. "Let's see it."

The avatar cocked its head, a smile beginning to crack. "You mean 'hear it.'"

"Probably."

"Anyhow, the battery is just a component. To retrofit the ships and maintain our BaaS solution, we need to be on location. Four of the biggest shipyards are in Korea."

"So the retrofitting and service operations will be in Korea?" asked Kasemchai.

The avatar nodded its head in agreement. "To begin with. Our first location will be in Ulsan. It's right next to the Hyundai, SME, and Samsung shipyards. From there, we could expand to China and Japan. More than eighty-five percent of the world's production is in Asia."

Kasemchai leaned back in his chair and stared at the ceiling. "So, the batteries are made in Germany, but the operations, the retrofitting, and support will be primarily in Asia." He rocked forward. "You're a genius, Liv."

"Thanks, Kasemchai."

Kasemchai slapped his hands on his desk and stood. "Let's celebrate. I'm taking you to dinner in Seoul."

"I'd love to, but I need to take care of some things. I'm sure I'll be out there soon. I do have a favor to ask, though."

"Name it."

"I need some introductions. We're going to be shipping a lot of batteries, and we'll need approval to open prime locations near the Korean shipyards."

Kasemchai smiled. "I'll figure it out. You leave that to me."

Chapter Thirty-Five

KASEMCHAI PULLED up the driveway of his parents' estate.

As he exited the car, Kosal hurried to meet him at the bottom of the stairs. He placed his palms together above his head and bowed deeply. "It's so good to see you back, Mr. Seng. Are you thirsty?"

"No, I'm not thirsty. Where is my father?"

"He's in the garden. May I tell him why you are here?"

"No, you may not, Kosal." A smile crossed Kasemchai's face. "Do I ever tell you why I'm here?"

"Of course not. You know, I am just trying to be helpful. You travel a lot these days, and your father is always so interested in where you've been."

They walked down the long hall through the great room and turned left after exiting the doors onto the balcony. Kasemchai knew that his father wouldn't be in the garden immediately below; that expansive area was for guests. His father's favorite location was his sanctuary. A more intimate garden, dense with more delicate flowers and carefully nurtured hanging vines. They crawled over statues of women with their hands pressed together—an homage to Angkor Wat. The temple was not behind the wall of rain but so close that it was never dry.

Kosal strode ahead of him and entered the garden first. He stopped and turned sideways. A small chime was produced from his pocket, and he tapped with a small stick. A melodic three-tone gong drifted along the garden.

Preap was holding a pair of shears, examining an orchid attached to the side of a tree. He turned and greeted his son with a warm smile. "Are you thirsty, Kasemchai?"

"No, I'm not thirsty, Father."

"Good, good. Please, come here. I want to know what you've been up to." He turned to Kosal. "Please, bring us some cool tea."

He turned back to the plant he was examining, a coelogyne pandurata, better known as the black-lipped orchid. Its leaves were creamy yellow, but the center was covered in rich, lacy black streaks that inspired its name. He leaned in to smell the fragrant honey scent. "Do you know the trick to growing a healthy plant is simply to let it grow?"

"Is that all, Father?"

Preap chuckled. "No, of course not. It needs to be trimmed back occasionally, it needs fertilizer, it needs the right amount of sunshine, and finally, it needs to be appreciated."

"I see," said Kasemchai.

His father stopped and regarded him. "What's new, Kasemchai?"

"Father, do you know about the battery company I'm investing in?"

His father put the shears down into his gardening box, removed his gloves and placed them over the shears. "I know you're working on making batteries for tanker ships."

"Is there anything you don't know?"

"It's important to know things. That's my advantage. But really—" He turned and winked at Kasemchai. "You're in the news all the time. How could I not know what you're working on?"

"So you're following Sol-vang."

"I am. A new type of battery for ships." He placed his hand on the small of Kasemchai's back and began to walk away from the garden. "Your PrefX was very impressive. So many people enjoy it. But people are fickle, and a PrefX is ethereal. People could simply stop using it one day. Building something material, that is something you should be proud of."

"Thank you. I am. In a way, I thought building the technology was the impossible part. But now I realize the really hard work is just beginning."

Preap smiled. "I'm glad you recognize that."

They followed a path that led them to a pool surrounded by a covered seating area. As they walked up the steps, they saw a table by the pool. It was just a foot off the ground and was surrounded by colorful cushions. Kosal had already placed their tea on the table, which was served in slender glasses with ice and fruit. The two sat down on the cushions, as they had many times before.

"We're contracting with a factory in Germany to make the batteries. But that's not what makes this such an interesting business. It's how we get the batteries into the ships. We're creating a BaaS model."

Preap carefully tilted a small carafe over his glass and poured liquid cane sugar into his iced tea. He stirred it with a long, slender spoon, clanging it against the glass. Without looking up, he asked, "A what?"

"Batteries-as-a-service. The shipping companies won't have to pay for all the batteries up front. They only pay a service fee. We maintain the batteries. As long as they're happy with the performance, we keep billing."

"Aha. You've found a way to mix your software thinking with a tangible business. That's a lot of support. Are you ready for that?"

Kasemchai took a long sip of his tea and then looked up at his father. "Yes. We're starting an industry. We have to build confidence, or what's the point?"

"I see. And you have backing?"

"I do. I pitched the idea early on to Mr. Sato. It was very challenging, but now he's on board. I know how this works, and I know that backers build momentum."

"Very good. You've thought of everything." Preap picked up a slice of dragon fruit and slurped it up. "And customers?"

"We're working on it. I've pitched the idea to a few groups, and they've shown interest. The retrofitting is a major capital expense, but the idea of running a ship with no bunker fuel whatsoever is very enticing."

Preap coughed into his hand. "I'm sorry, no fuel? I thought this idea was to save on fuel."

Kasemchai laughed, but his eyes were serious. "I didn't mean to make you choke." His father waved away his concern, so Kasemchai continued. "It's bigger than that. We completely remove the engine and replace it with batteries. No fuel. No Saudis."

"I see." Preap pursed his lips and nodded up and down slightly. "Hmm. That is interesting. You say you can completely remove the engine and run entirely on batteries?"

"Absolutely," Kasemchai said confidently. It was far easier to sound confident than to understand how this technology would work, but he knew that tone was far more persuasive than facts.

Kosal walked down the corridor towards them. He approached the table and kneeled next to it. "Would you like anything to eat?"

"Yes. Bring us some sach ko ang," Preap said. "And wine. I must keep this conversation going."

"Very good."

"Kosal."

"Yes, sir."

"Do you use the Serendipity PrefX?"

Kosal didn't look at Preap, but a tiny smile could be detected on his face. "I do, sir. Thank you for asking about my personal life."

Preap waved his hand dismissively. "No, no, no. I don't need to know that. I was just wondering if it's clever."

"Very much, sir."

"Okay, okay, we don't need to get too personal. I just received a new case of Cabernet. Would you open one of those bottles?"

"Of course, sir."

The uncomfortable conversation with Kosal made Kasemchai laugh, but he kept it to himself. He turned to look at the blue lights in the pool and was surprised to realize that the sun had set. The conversation with his father was going extremely well.

Kosal returned with the wine and two glasses. Kasemchai tried to reach over and pour the wine, but his father swatted his hand down.

"I'm far too young for you to serve me."

Kasemchai laughed. The red wine swirled about the large bowl of his glass. The men toasted, swirled, and tipped the ruby liquid into their mouths. Kasemchai detected strong notes of plum and black cherry. He sighed contentedly and placed the glass down.

"I think you are intrigued by the idea of shipping water without paying outrageous fuel fees to the Saudis," Kasemchai said. "Extra profit or reduced costs for consumers—either way, you come out ahead."

Preap shook his head from side to side. "Always. My mission is to keep costs lower. Water is not a luxury. It is a right. My success is built upon my obligation to bring water to the world as cheaply as possible."

Kasemchai shook his head in agreement. "Of course. You've told this to me before. I shouldn't have phrased it that way."

"And profit—" He swung his arm wide, just missing his glass of wine. "You're not thinking of the cost of retrofitting, new feasibility studies. Will the ships be fast enough? Will any ships fail at sea? No, change never brings new profit."

Kasemchai's smile faded. "Perhaps we should consider picking up the cost of the retrofit, at least until the model is proven."

"This is not a car engine, Kasemchai. This is a huge undertaking. The engine room is as big as a house, and it's filled with toxic chemicals, oils, and asbestos. Where will you put the engine once you've taken it out? What will happen if a customer wants to put their engine back in?"

Kasemchai looked down at his wine. He rested two fingers on the smooth foot of the glass and slid it around on the table.

"Don't worry, my son. You will have to answer these questions from much harder customers than me," his father said. "But I believe you have an interesting idea. And I know you're working with smart people."

"Thank you. I will do the hard work." His gaze darted out to the pool, and then back to his father. "I'd like to know something."

"Yes."

"Father, would you consider changing the engines in your ships for batteries?"

"No."

He was caught off guard by the simplicity and finality of his father's answer. His jaw clenched. He tried to show no emotion, but a single word betrayed him. "No?"

"Do you know why the rain falls only in Cambodia and Africa?"

Kasemchai's mind raced. His father, as always, wanted to make this into a lesson. Maybe if he kept talking, he might be able to change his mind. "Well, there are theories . . ."

"Yes, there are lots of theories. But the most prevalent one is pollution. Long before the rain fell in two places, we had experienced unbearable heat. Trees were already dried up and burning. Winter was nothing but a phrase in songs." He pointed over the pool towards his garden. "What do you see over there?"

Kasemchai stared over the pool. "A garden?"

"Yes, a vibrant garden. The glorious, colorful plants you see are a treasure for us, a reward for us, and a mystery."

Kasemchai laughed cautiously. "Why is it a mystery? You have all the water in the world. Of course, your garden is green."

"It is a mystery because someday, the world could be filled with gardens again." He looked over his garden and turned back to Kasemchai, but there wasn't joy on his face. "Or we may lose this garden because we have lost our Rain Wall."

Kasemchai stared into his father's eyes. "What are you saying?"

"I don't know what the tipping point will be. Will it be the last cup of water a child drinks? One more car on the road?" Preap placed his fingers on the base of the glass and dragged it toward him, but he didn't drink. He only stared at the glass. "All I know is that the world needs to stay the way it is."

"Are you saying that you want to ensure we continue to burn oil? As much as you hate buying fuel, the thing you fear most is not buying fuel?"

His father snorted but said nothing.

"And what about Philippe?" asked Kasemchai.

"Who knows? We both know what brought us our fortunes." His eyes twinkled and he grinned. "But he's also greedy, so who knows."

"And if he retrofits some of his ships, would you reconsider?"

"We'll see. But don't forget, Kasemchai. It is your duty to protect your family, not change the world."

Kasemchai lowered his eyes and stared for some time at his glass. He was so deep in thought that he was startled when Kosal knelt beside him.

Kosal placed the sach ko ang on the table—long bamboo skewers with sizzling cubes of beef. The fragrant yellow Kroeung paste filled Kasemchai's nostrils with the scent of lemongrass and ginger, a reminder of happier days.

Two women knelt beside Kosal and laid more plates on the table. A blue ceramic platter filled with juicy oysters, steamy saffron-colored dumplings pulled up into twisted points, fruit delicately cut and arranged into flowers and animals, thick foie gras lobes with diamond char marks, and bowls filled with toasted baguettes, brioche slices, and five different kinds of pickles.

His father slurped down an oyster and grinned at his son. "You have a PrefX that connects lovers, and yet, I still have no grandkids. How is this possible, Kasemchai?"

Chapter Thirty-Six

THE SERVER LAID down a plate in front of Liv. The smooth cerulean blue ceramic contrasted with the almond-colored parchment paper. Three large tempura spiders, dusted with red paprika, were neatly arranged beside a glass dish filled with sweet and spicy chili sauce and three grilled green onions.

Liv smiled at her dish and looked up at the server. "I love his arrangements. Each plate is like a work of art."

"Thank you. The chef believes all the senses must be entertained. Long before you have your first bite, your brain has engaged with the sight and the scent." He gave a sharp, quick bow. "Would you like anything else, ma'am?"

"No, I'm good." Liv picked up a spider, dipped the legs into the sauce, and took her first bite. "Mmm."

Her AugWatch hummed on her wrist, and she glanced down to see that Kasemchai was calling her. "Hello?"

Kasemchai's avatar appeared, seated across the table from her.

"Hey, Liv. How's the pilot going with the Fuchs personnel carrier?" asked Kasemchai.

Liv dropped her spider onto the plate, breaking off one of the delicate legs. "This is not the place for that conversation, Kasemchai," she hissed. "I'm in a restaurant."

"Then go outside. We need to talk."

Liv looked up at the server and pointed with her index finger. The universal symbol for *I will be outside for one minute and will return afterward*. He seemed to understand.

Liv made her way to the front of the restaurant. She stepped through the door, which closed on Kasemchai's avatar. It flickered for a moment and then appeared to walk right through the door and follow her outside.

She scanned her surroundings and picked up the conversation again. "You must understand by now that I can't talk about a confidential government project in a restaurant."

"Of course. Anyway, are we actually running tests yet? I need some solid news."

Liv rolled her eyes. She was now used to the way he crashed forward when he had an idea. Rough as they were, he had good ideas more often than she liked to admit. "Well, actually, the tests are going quite well. We've increased the range since our earlier tests by twenty-three percent."

"Good. And testing in water?"

"No issues. The batteries are completely inert, so we really didn't expect any impact on them. The only point for corrosion is the outside case. No issues."

"Okay, so we're testing in seawater, and no issues."

Despite Kasemchai's avatar in front of her, Liv was suddenly tired of the visual charade. She stared at her AugWatch, the source of her apprehension. "None. Why?"

"And no updates on the ships?"

"No, the pilot is running well, but still no contracts with shipping companies."

"Okay. Here's what's happening. I'm working on a proposal for the Labonne family. I don't have a formal meeting yet, but I'm working on it. And I'll need you in the meeting."

"Of course, but when are we presenting?"

"I'll let you know when I have more details."

He ended the call.

Under her breath, Liv muttered, "No goodbye. Why does he do that?"

She turned and marched back into the restaurant to finish her meal.

Chapter Thirty-Seven

KASEMCHAI'S BENTLEY had four leather reclining seats. The self-driving car's cabin was so enormous that the seats could be reclined to twenty-eight degrees and the footrests fully extended. He kept his seat back in a somewhat upright position so that he didn't fall asleep immediately. He drove west, the Rain Wall receding behind him.

"Hey, Chanthavy, call Angélique."

After a few audible tones, Angélique's avatar appeared, sitting in the seat across from him. She smiled. "Hey, it's been a long time, Kasemchai."

"Did I catch you at lunch?"

"You did. I'm impressed. The amazing Kasemchai actually paid attention to the time zone."

Kasemchai chuckled. "Yes, it's true. Even I can become more mature. How're things?"

"They're good. The family is well."

"Uh-huh, that's good," Kasemchai said. He knew he was stalling. "I'm planning a trip out there. Maybe you and I could have lunch?"

"That would be nice."

There was a very long pause.

Kasemchai tried to break the silence. "I could—"

Angélique started simultaneously, "You could meet my fiancé."

Kasemchai answered before his brain could process this news, "What? I mean, that's fantastic." He pressed the button on his armrest that straightened his reclined seat to almost vertical. "I'm so happy for you. Who's the lucky guy?"

Kasemchai had set the windows to transparent mode and the trees and houses to the right of Angélique sped by. The ocean on the left was expansive, never changing. On that side, it appeared they weren't moving at all.

"Faya Camara. You know, now that I think about it, we met not long after you had your trip to Ishigaki Jima."

Kasemchai had a sour look on his face. "I've heard of him. That's great, Angélique. When is the wedding?"

"Next winter. We didn't want it to become a big media fest, so it will be held at the Crystal Egg. I'm surprised your father hasn't told you?"

He stared up at the ceiling, confused. He didn't see her that often. They never dated. But no one in the world understood his station in life better than she did. And the sex was good. It felt . . . forbidden. No, maybe the word he was looking for was comfortable.

He knew the right thing to say, but he couldn't stop his mouth. "Does he love you?"

Angélique's avatar locked its eyes on him. "Did you just ask if my fiancé loves me?"

Kasemchai threw his arms up in defense. "I'm sorry, I'm sorry. I don't know why I said that. I'm so busy with this new company. I guess I haven't been paying attention to the rest of the world." He lowered his hands and said with genuine sincerity, "I'm really happy for you, Angélique."

"Thanks. It's a good thing you have other redeeming qualities, Kasemchai." Her tone softened. "So, tell me, when are you coming out?"

"I don't have an exact date yet. I have a new company I'm funding, Sol-vang. Have you heard of it?"

"No, what kind of PrefX is this?" she laughed. "I just told you I'm already engaged."

He realized he might never get away from the dating moniker, and his voice rose in pitch in defense. "No, no, no. This isn't a PrefX." He brought his voice down to normal so he didn't sound like a whiner. "This is real. Tangible. I'm investing in a battery company. We're going to revolutionize the shipping industry."

Angélique's avatar stared at him with its mouth open. "That's unexpected."

"Yes. This venture will have a real impact on the environment. And it could break the dependency your family and mine have on oil. We could significantly reduce the cost of shipping water."

"So, the golden child has suddenly realized our fathers are running a business." A slow grin spread across the avatar's face.

He winced. "I have my own business interests, Angélique. I've just never paid attention to their business until now."

"So why are you coming out?"

"I need to set up a meeting with your father."

"Why talk to Papa? Surely, it would be easier to twist the arm of your own father first. Or has he already turned you down?"

Kasemchai hesitated for only a second, but he was prepared for this question. "The thing is, your father ships far more water than we ever will."

The eyes of the avatar narrowed, and its jaw shifted slightly. Angélique responded with a measured tone, "You say my father, but I like to think of it as a family operation now."

"Of course. Your family ships far more water than ours will. My father wants to lower costs as much as you do, but he's getting soft. I know Philippe wants to continue to grow aggressively, and this could give him a major advantage."

"You mean give *us* a major advantage."

Kasemchai shifted in his seat. She was goading him. She always found a reason to twist the knife, even when they weren't competing. "Of course, it gives you a major advantage. It sounds like you're more in the business than I realized."

Angélique's avatar leaned on the armrest toward Kasemchai. It held up its hand and clicked its glossy thumbnail slowly and deliberately against the tip of the nail on her middle finger. "You would be wise to do the same, Kasemchai. Life is transient, and so is wealth." It turned its hand over and let it hang limply in the air. "You know you could just call him, Kasemchai. You're like family."

"I know, I know. But I don't want him to think it's just a family favor. I have a serious proposal I want to run by him."

"Okay."

Kasemchai began to question his plan. Of course he could call Philippe directly. His mind raced as he began to think about how he could work around her. He realized it would be foolish now that he had already broached the subject. "If you could just drop some hints about the business, I'll send you a few articles. Get his mind thinking about the high price of oil and what could be different."

"Don't worry about it, Kasemchai. I'll set up a meeting with our family."

"Great, thanks. And Angélique."

"Yes."

"I really am happy for you. You deserve this."

The avatar reached over to Kasemchai's hand. It hovered just over his as if it meant to touch him and changed its mind or was distracted. "Thank you, Kasemchai. I'm glad you feel that way."

Kasemchai ended the call. He stared at the AugWatch for a long time. He didn't understand why he cared that she was engaged, but he felt he had lost something—like a door had just closed on him.

Was it an opportunity with her that he could no longer pursue? Or maybe it was him. He stood still while people around him moved forward.

He slammed his fist on the armrest. "Chanthavy, call Annop."

Chapter Thirty-Eight

KASEMCHAI WALKED QUIETLY DOWN the hall towards the aquadrome entrance. At the far end, the attendant opened the door for him. He started toward the back of the room, and his face immediately broke into a smile when he looked up to see his friends in the center alcove.

Annop stood and raised two martini glasses. "Got your Gibson right here, Kasemchai!"

The four friends stood around the cocktail table, the music pounding in the auditorium. They raised their glasses, clinked them together, and then drank deeply. The cool gin spilled down the sides. He gingerly laid the glass down and shook the alcohol off his hand.

"Spilling, already," Kasemchai boisterously said. "This is looking like a good night, boys."

When they sat down, Annop leaned into Kasemchai. "You all right, man? You sounded like you needed some serious cheering up."

"Yeah." Kasemchai took another swig from his Gibson. He shouted over the music, "You know how sometimes you think you can know someone, and it turns out to be exactly the opposite?"

"I'm sorry to hear that. Someone I know?"

Kasemchai smiled. "Maybe." He stared at the translucent pearl onions submerged in his glass for a long moment, then turned to his friend. "You know what? It's just business. You have to remember that it's not about how you change the world—"

The two raised their glasses and toasted, spilling more of their cocktails. "It's power and wealth!"

Kasemchai swept his glass around to toast the rest of the group. "Power and wealth, my friends!"

The conversation grew louder and louder with each drink. They laughed harder at their jokes and shared stories of how to spend their money and the weaknesses of others. When the mermaid and aquatic dancers entered the pool, the boys only momentarily stopped to stare.

The room darkened, the buzzers went off, and the underwater rugby divers dove into the pool. The men stood and cheered, slapping backs and punching their fists in the air. The pool looked like it was filled with eels. Men and women were fighting over the ball, turning round and round, chasing the ball up, and then dolphin kicking down after it as the ball shot towards the goal.

Kasemchai was rooting for Club Chang. Every time the ball was on the Tigers' side, two defenders blocked the goal by sticking their heads and shoulders into the basket and furiously kicking their legs to keep their bodies downward. As soon as one shot up to the top for air, the other shot downward to take her place.

Legs, arms, and bodies pushed and undulated. They fiercely grabbed at the ball and pulled on each other. Quickly rising to the surface to grab air, they shot back down to join the action.

Club Chang was falling behind, and Annop rejoiced in the other's misery.

Kasemchai stared up at the ceiling, and a mischievous grin broke out on his face. He turned and called over their attendant. "Tell the Club Chang captain that if they win, everyone who played gets a Mercedes EQ Opal."

The attendant stared at him, eyes unblinking. There were deep furrows in his brow, and his mouth opened and closed again.

Kasemchai put his hand on the man's shoulder and leaned forward. "Did you hear me? Everyone gets an Opal. They just better damn win. Go on, tell him."

The attendant scurried off.

Kasemchai turned back to Annop. "How confident are you that they're going to win, my friend? You sticking with your wager?"

The two bantered back and forth. The stakes were raised. Annop looked pleased with himself until he noticed how much more aggressive the players became. Club Chang was animated. They played faster, harder, almost violent.

The play stopped twice for players with bloody noses—a rare occurrence in the sport, but tonight was different. In the last minute, Club Chang was ahead by two points. The Tigers had the ball and fought furiously to score a last point.

A Club Chang player climbed over his shoulder and stole the ball. She threw it back to her opponent, who broke away and swam furiously towards their basket. Just on top of the basket, several players from both sides wrapped around him and spun around and around.

A woman grabbed the ball. He pulled back. They locked eyes, and he realized she was his teammate. He passed the ball to her, and she slammed it into the basket and shot to the surface.

The buzzer sounded, and she broke through the surface and yelled in triumph. She slapped the water, and it splashed out over the crowd below. People were on their feet, screaming and jumping up and down. Even those far from the pool were wet from the drinks that seemed to explode from their glasses.

Kasemchai roared at the top of his lungs, and the crowd noise enveloped him. He kept yelling until he knew his voice was cracking, but even then, he couldn't stop. When only strangled croaking sounds came from his mouth, he shook his hands and danced around wildly. He was enveloped with joy.

Chapter Thirty-Nine

THE OFFICE WAS STILL. The sun was just beginning to rise, blood-red with wisps of light orange. The smoky air blocked all the shorter blue and green wavelengths.

Liv gazed at the sunrise until the electrochromic window tint grew so dark that she could no longer see outside. Everyone had gone home long ago. She had spent the time reviewing the research from their pilot program, including the cost analysis she had conducted on switching to batteries, the convenience of the battery-as-a-service solution, and even the time it would take to retrofit the ships.

She struggled with the timeline for the retrofitting process. It was too long and the argument for switching to batteries wasn't compelling enough. If Philippe grew impatient with her presentation, she feared she would lose the opportunity. She closed her presentation.

"Noa, I'd like to write notes."

She placed her thumb and index finger together on her desk, then pulled them apart to form a rectangular shape. The blank white canvas sat before her as she picked up her digital pencil and began jotting down notes. She wrote down fuel disadvantages and listed them:

- Pollution - high levels of sulfur dioxide, carbon dioxide, black carbon particulate matter
- Availability - oil is a finite resource - long-term liability
- Budget - volatile pricing - difficult to manage fuel budgets
- Slow-steaming ships run below max to reduce emissions and fuel consumption (24 knots to 19)
- Maintenance of complex engines - constant maintenance and highly trained crew to maintain
- Idling time - heat exchanger preheats bunker fuel before it reaches the engine (hours to days)
- Ballast - weight loss from fuel/water consumption = pumping sea water into ballast tanks (Panamax consumes 63,000 gallons of fuel per day)

She examined the list. "Shit, what a ton of work to manage these things. It's amazing no one thought of switching earlier."

She jotted down battery advantages:

- No pollution, zero emissions (charging or underway)
- Noise pollution + energy efficient looped propellers - ships have virtually no noise and are 30% more efficient
- Budget - lower operating costs in the long run
- Idling time is essentially none
- Engine maintenance largely displaced by BaaS
- Crew reduction - chief engineer and electrical engineer remain. First assistant engineer, second assistant engineer, third assistant engineer and motormen/engine ratings positions eliminated
- Instant response - nearly instantaneous power delivery - quick adjustments to propulsion and stabilization systems
- Regenerative braking sends electric energy to batteries and offsets the ballast needs of bunker fuel

- Reduced weight - batteries lighter than the equivalent amount fuel/reduced ballast water required for ship stability

She looked up from her desk and sighed. "Okay, Noa, using my advantages and disadvantages, give me a persuasive presentation for a company with fuel-based tanker ships to switch to batteries."

She watched as the notes appeared. Noa outlined it with *Slide 1: Title* and several bullet points, followed by *Slide 2: Introduction* and more bullet points. Continuing until *Slide 13: Conclusion.*

She studied the outline and then said, "Modify 'battery' to 'Battery-as-a-Service.'"

She scanned it again, then said, "Make it emotionally persuasive."

After reading the changes, she realized it had lost its edge. "Okay, but focus less on the environment and more on the financial benefits."

She read it one last time. "I like where we're getting, Noa. Could you tell this in the form of a story instead of a presentation?"

Noa created an outline with the title: "A Voyage to Financial Triumph: The BaaS Transformation" followed by a short paragraph. Then, *Act 1: The Financial Storm*, followed by another paragraph. Then, *Act 2: Discovering the BaaS Treasure.*

She laughed. "Noa, this is starting to sound like a story about pirates. Tell the same story, but don't use so many sea metaphors. Get right to the point. And make it shorter."

She sat back and watched the words scroll in front of her. A big smile broke out on her face as she watched the story unfold. "Noa, I love it!"

"Thank you, Liv. Have you thought about what the presentation will be like? Will it be in a conference room or more intimate?" asked Noa.

She absentmindedly tapped the digital pencil against her index finger. "I honestly don't know," Liv said.

"Well, I suggest some images to go with your presentation."

She stopped tapping. "Absolutely. What do you suggest?"

"How about a before-and-after comparison showing an image of the ships with engines and ballast tanks and then replaced with battery compartments? This could show how the ship would gain additional storage."

"That's excellent."

"You should also show images of the battery technology, financial charts showing reduced energy needs, predictable maintenance budgets, and rising profit margins."

A chart hovered over her desk with an arrow on the side. With her finger in the air, she swiped through several chart options before asking, "Do we have to show this as a chart?"

"Of course not. A chart is a format people are familiar with for comparing information. There are many ways to compare information, and there are many ways to show numbers. I believe what you are asking for is a more relatable way to express this information. Is that correct?"

"You got it."

"Of course. Additionally, people have a stronger emotional connection with images of people, especially when they can see their faces. It's a good idea to include images and videos of people adversely affected by pollution and how they would feel after the conversion when they receive clean water from clean energy ships."

She turned and stared at the window, then realized she was staring at her reflection. She grinned and turned away from the window. "I like that."

"One final suggestion, Liv. While you think the environmental impact of this solution is important, Kasemchai is mostly focused on the benefit of ending the dependency on oil. He's specifically focused on Saudi Arabia."

"Sure. Where are you going with this?"

"It borders on an obsession for Kasemchai."

Liv sighed. "I know. I get it, Noa. What's your point?"

"There is a high probability that Angélique will be in the negotiation with you," Noa said.

Liv dropped her gaze down to her AugWatch. "Why do you say that?"

"Kasemchai's obsession is really not with the Saudis. It's closely aligned with something of importance to Angélique. They are both concerned with how they fit in the future of the two water families. And they are both concerned with their parents' legacy and how they fit into that story."

Liv cocked her head to the side. She ran her tongue between her teeth and gums. "I have to say I'm blown away, Noa. I was aware of Kasemchai's obsession with the Saudis, but Angélique . . . that's news to me. How do you know this?"

"I don't have any more information about Angélique than you do. However, Kasemchai uses phrases and mannerisms when talking about his family and Angélique. I'm able to build a behavioral profile based on cues that are not observable at a surface level."

"Fantastic. Let's use that."

"Okay, I made a few adjustments. Would you like to practice your presentation now? I can pretend to be your audience."

"That would be great."

Avatars for Philippe and Angélique appeared in the chairs across from her desk. Philippe stood up and spread his arms in a welcoming manner. He sat down again and said, "Liv and Kasemchai, thank you for coming to visit us today."

Liv scanned over her notes so erratically that it was just a jumble of ideas in her head. "Thank you, sir. In just a few minutes, I'll show you how the cost of battery swapping will make your time in port shorter by eliminating engine maintenance." She stopped. "Noa, tell me a joke."

"Okay, uh, you know what can ruin your Friday?"

"What?"

"Remembering it's Thursday."

Liv stared off into the corner of the room. "Do you think I should start with a joke?"

"Sure, Liv. But it's important to understand your audience, keep it light, and keep it relevant. That was not a good joke for this presentation."

Liv grabbed her head and let out an anguished cry. "I can't do it! There's too much riding on a single meeting. If we don't get Philippe to sign, we're finished." She clasped her fingers over her head and stared up at the ceiling. After a moment, she let out a deep and tortuous exhale. "Noa, show me how you would start this presentation."

"Sure, Liv. Why don't you sit where Angélique is sitting? We'll have Angélique give the presentation."

Liv strode around the desk to sit down next to Philippe's avatar, and Angélique's avatar stood up and walked to where she had been standing. The room around Liv darkened, and the scene behind Angélique was underwater. She could see a giant propeller turning slowly, bubbles streaming behind it. Then they broke up through the water, and she stared up at the enormous, rusted steel façade of the ship, slowly panning up and towards the bow.

"Imagine the vastness of our oceans, where massive tanker ships traverse the waters, connecting nations and economies. These ships are colossal engineering marvels that have long symbolized the might of global trade. But they've also been symbols of something else—a challenge many believed was insurmountable."

The view behind Angélique flew out over the ocean, away from the ship. Greasy smoke streamed from behind the bridge, exiting a tiny port and growing in size as it dissipated into the air.

"For decades, these giants of the sea have been powered by fossil fuels, belching emissions into our skies and leaving a carbon footprint that weighs heavily on our conscience." Angélique's avatar held up her hands like a boxer and clenched its fists tightly. "They seemed like immovable forces of nature, impervious to change. The conventional wisdom was that something so large, so heavy, could never run on anything but the power of diesel or heavy oil."

The background panned to the sun until it obliterated all detail. There was nothing but white. And then she could make out that they were falling back to Earth. A spec became a large outdoor solar panel installation.

"But today, I want to share with you a different narrative—a heroic narrative. It's a story of transformation, of defying the odds, and of rewriting the script for our future." The avatar held its hand up, pointed to the sky and shook its fist. "It's the story of how these seemingly unchangeable vessels are transitioning to something cleaner, greener, and more sustainable."

Liv could see large cranes dropping batteries onto a tanker ship. They zoomed in, and she saw the engine room was stripped clear of the engine. The dirt and grime disappeared. It was replaced with new paint and clean racks for the batteries to be loaded onto.

"This is not just a technological shift; it's a revolution—a revolution on the high seas," Angélique's avatar said. "It's about daring to challenge the status quo, to envision a world where these leviathans of trade and commerce are no longer agents of environmental harm but champions of sustainability. It's about the audacious idea that something as massive and powerful as a tanker ship can be tamed, can be transformed, and can run on batteries."

They flew onto the bridge. The captain's hand moved across the controls and tapped a button. His face was wizened with deep crow's feet, and he stared off into the distance. They zoomed over to the bow to see the endless ocean in front of them. Then they flew over the railing, over the sea. Liv looked down from a dizzying height, foamy water crashing into the bow.

Angélique's avatar looked down at her and smiled. They locked eyes for a moment, then Liv released a small gasp as they dropped to the sea. Just before they smacked the ocean, they looked out over the horizon again—the wall of the ship to her left. Bits of spray whipped up from time to time. She was lulled by the rhythm of the ocean slipping beneath her.

Angélique's avatar walked slowly and deliberately around the desk to stand in front of Liv. Liv watched, mesmerized. "Ladies and gentlemen, I invite you to join me on this heroic journey. Together, we will explore the possibilities, innovations, and uncharted waters of a new era in shipping—a future where the largest ships afloat not only carry cargo but also the promise of a cleaner, more sustainable world. It's a narrative of courage, innovation, and a commitment to leave a lasting legacy for generations to come."

Liv returned her gaze to the avatar for Angélique. She began to clap slowly, saying only one word: "Woah."

Then, she clapped vigorously.

Chapter Forty

MAKOA's AugWatch woke him up at 2:30 p.m. He groaned and reluctantly got out of bed, knowing that today marked the beginning of his new life as a tankerman. After pulling on his stiff coveralls, he lifted the shutter to look out the window. It was daylight, and the sun wouldn't set for five hours. He thought to himself, *No one should be up and outside at this time of day.*

He had completed two months of tankerman training and now walked down the narrow gangway to the bunker barge he would crew on. Even with the motley shade structure, the metal surface of the deck radiated heat.

The deck officer, Ben, greeted him with a gruff nod. He removed his glove and shook Makoa's hand. "Welcome to the dirty world of tankermen, rookie. Don't screw up, and you'll walk that gangway down here again."

Ben's face was dark and shiny, like the fuel inside the bunker barge. Thick creases crossed his face in haphazard directions. They weren't laugh lines; he was just weathered. He stooped a little with age but still appeared strong.

Makoa looked backward at the gangway that connected him to the mainland. It rose and fell in a sickly move with the ebb and flow of the bay. The crackle of the captain's voice through a speaker pulled him back to the task at hand. "Don't let me see you not wearing your gloves again, rookie."

A dry laugh escaped the deck officer as he put his glove back on. "You know we work with a lot of chemicals, and the hoses are hot as hell. Keep your gloves on at all times. You get injured, and I have to find a replacement."

Ben introduced Makoa to the other deckhands. The unmistakable scent of bunker fuel filled the air. It was a pungent mix of oil, saltwater, and exhaust fumes that never quite left your nostrils. Their job was to refuel the massive tanker ships that loomed nearby.

As they approached the first ship, Ben directed the deckhands to catch the mooring lines. The barge slowly pressed against the tanker ship and rocked back, straining the lines before drawing forward to the tanker. A boom on the bunker barge was lowered, and they connected a thick black hose. It was the diameter of a palm tree and required two men to lift the heavy metal end.

The deckhands and engineer boarded the ship and bolted the hose to the manifold. They took soundings of the tank, checked all the cleaning equipment and safety protocols were in place, and confirmed the discharge. The viscous bunker fuel began to pump into the ship, causing the hose to buck and slap down against the railing like a trapped anaconda. A tanker ship of this size would consume between a hundred thousand to two hundred thousand liters per day. They filled this ship with just over nine hundred thousand liters.

When finished, the tanker crew cast off the lines. Far below, the crew on the bunker barge cleaned up the deck and stowed all the gear. Makoa walked to the aft and stared down at the sea. It turned frothy as the engine cranked on, thrusting the water out in a long wake as it left the tanker behind.

The ocean always calmed him, but the smell of exhaust drifting up from the broken sea made his head feel like it was trapped in a vise. The sound of the engine was deafening, and the deck trembled beneath his feet.

The job was dirty, miserable, and repetitive. This was what he would do every day for eternity. His heart raced with the thought that this was his punishment. He deserved this, but he knew his two little girls deserved better.

Chapter Forty-One

KASEMCHAI STOOD with Liv under the hotel's overhang. He looked up as three identical cars approached. "Here we go."

He gave her a quick wink. He was nervous, but the look on her face was more like terror. "Nothing to worry about, Liv. I've known this guy for years. He used to come over to our home, and my father and he would spend hours in the pool, just getting drunk and splashing water everywhere. He's just a man."

Liv glanced up at Kasemchai. "A, no man I know has ever sat in a pool splashing water. And B, I'm not nervous. That's just my thinking face."

"Really? Because you look like you're about to be sick."

"That's what I look like when I think about something that is illogical . . ."

"Oh, my god, Liv. Was that a joke?"

Kasemchai attempted to stand still as the car approached when he was overtaken with giggles. He tried to stifle the sound through his pursed lips. He turned to look at Liv, who was laughing with him, through blurry eyes.

"Try to pull it together," he said. But that just set off another round of giggles.

As the car slowed in front of them, he wiped his eyes clean. The middle car came to a stop. Men from the front and rear cars ran to the middle car.

The car's doors opened automatically while the two men stood stiffly on either side—one door opened towards the back, the other towards the front. The roof lifted up and extended out, providing both a cover from the elements and headroom.

Angélique stood and walked towards the edge of the car. One of the men held his hand out for her as she stepped forward and out of the automobile. She had arrived, and Kasemchai noticed it. This was not the girl he had been intimate with. She seemed more mature. In fact, like the long silver spike of her heel, dangerous.

She strode up to them, opened her arms out with the palms up, and bowed just slightly. "Are you thirsty?"

Kasemchai and Liv answered, "No, I'm not thirsty."

"It's good to see you, Kasemchai. And Ms. Anselm. Welcome to Côte d'Ivoire." Her French pronunciation chopped the sound of "Côte" and rolled the "r" in "d'Ivoire." Kasemchai noticed how different it sounded from the more rhythmic and fluid pronunciation he had heard from the locals.

Angélique shook Liv's hand and gave Kasemchai a distant hug before turning to enter her car. Liv and Kasemchai followed her, taking seats opposite Angélique.

As the car accelerated, Kasemchai leaned forward. "Thank you for picking us up. Should I start calling you Mrs. Camara?"

She smiled with pursed lips. "You can just call me Angélique."

"Are you nervous? I mean, about the wedding."

"No. We love and respect each other. That's all that matters." She looked over at Liv and back to Kasemchai again, then raised one eyebrow. "Tell me, how much longer will Kasemchai be a bachelor?"

Liv turned away to avoid the verbal jousting between the two.

Kasemchai shrugged. "It's a lifestyle. No one has been able to pin me down yet."

"Hmm. And you, Liv. There's a lot about you in the news recently."

Liv turned back to Angélique. "Are you referring to our contract with the Bundeswehr?"

"I am," responded Angélique.

"People are quite excited about electric military vehicles." Liv leaned in, propping her elbows on the armrests. "It's amazing the military hadn't solved this earlier. Stealth is a strategic advantage."

"But that's just it. No one has done it before. That accomplishment is no smaller because it's behind you." Angélique took a drink of water from the crystal glass beside her. She carefully placed the glass down before regarding Kasemchai again. "And I still can't believe you're behind this. You have changed." She faced Liv and winked. "Although we know that Liv is the brains."

Liv leaned back in her seat again, a tinge of red touching her cheeks.

"She is, indeed." Kasemchai chuckled. "Now we just need a fleet to prove our model at scale."

"Yes, of course, Kasemchai. We all know why you're here today. Our family is keen to innovate, but you know father is a methodical man."

"I was curious about that," Liv's eyes lit up, and her energy noticeably spiked as she steered the conversation back to her passion. "When I watch him in virtumersion, he seems so gentle. But I hear he's quite the passionate engineer."

"That is correct. He is grateful for what God has bestowed on him, which is why he devotes so much time to the ministry. But as he often says, 'The devil is in the details.' He feels strongly that God rewards those who work the hardest."

"So, does he see the batteries as an intellectual challenge?" asked Liv.

Angélique laughed. "I think the reason we are here may surprise you. But at the end of the day, for Papa, it's oil. He can't stand that it keeps rising from the ground, but the price never goes down."

Kasemchai slapped the armrest. "Hah! He and Father are no different!"

Angélique let her smile begin but then held it back.

Liv looked out the window the rest of the way. She marveled at the lush, green jungle. The rain tapping on the roof of the car startled her at first but then became soothing as she felt the humid air invade the cabin.

Angélique pointed out the window when the car emerged from the jungle. The Crystal Egg appeared in the distance. It emanated light and seemed to pulse with energy from the waves of rain.

Liv stared, her mouth agape as her eyes darted in every direction. "Pictures don't do it justice."

"No. I know that rain must seem like a luxury to you. Here, it never stops. But Papa insisted that he live in the rain. He says it's the source of his energy."

Liv saw an enormous parrot and pointed at it. "How do the birds fly in the rain?"

"Birds have always flown in the rain. But we are still in the outer ring. Deeper inside, there are animals you have never seen before. The further you go in, the darker it gets. Sponges, and water ferns. Creatures that can't decide if they are amphibians or fish. It's hard to say if you are on the water, in it, or under it."

Kasemchai laughed. "So, you do use my PrefX."

"Only to keep tabs on you, my friend."

"I heard there are people who live in there?" asked Liv.

Angélique looked up to see what Liv was studying. "It's possible. In truth, we don't need to go into the Rain Wall. We only need to build dams and aqueducts on the outer ring to collect the water."

The car pulled under an enormous glass overhang, more than fifty feet above the driveway. The three exited the car and walked up the stairs, past a long line of servants, and through the great doors. Angélique led them through the atrium toward a glass-walled column.

Liv stared, watching birds gently sweep from branch to branch high above her—enormous trees with thick cords of vines draped down to the ground. She was so captivated that she bumped right into Kasemchai's back. "I can't believe this place. I - I can't believe I'm here. I've never seen so many shapes and colors in my life. And look at those leaves. They're as big as a human."

Kasemchai turned and smiled. "I've been here a few times. It is breathtaking each time."

They stepped into the glass elevator, and Liv darted to the glass wall. It shot upward and then settled into a steady acceleration. She let out a little gasp. Her knuckles whitened on the railing. Her mouth hung open as she scanned the canopy.

Kasemchai gazed out over the garden, but he couldn't shake a thought. Angélique was different towards him. Maybe it was her engagement. Or maybe because he was there on business. Whatever it was, she was very distant and measured.

The elevator comfortably decelerated, and the doors opened. Kasemchai looked down at Liv. "You can close your mouth now."

Her mouth snapped shut, and her earlobes tinged red.

They entered Philippe's office and walked towards a long table. Kasemchai had never been in here before. He imagined his father and Philippe meeting here, deciding the destiny of the world. He was finally in the inner sanctum, pitching Philippe on his own.

Angélique sat on one side and gestured for Kasemchai and Liv to sit opposite her. Two attendants entered and placed a tray with a carafe of water and glasses on the table. They also placed a wooden tray with wedges of cheese, several different types of crackers, and a small bowl filled with jam.

Liv looked hard at the tray.

"Would you like some?" asked Angélique.

"Perhaps. Is that . . .?"

"It's cheese. My favorite is the smoked Gouda."

"Cheese? You mean from a cow?"

"Yes, Liv. I suppose there's goat cheese. But most cheese comes from cows."

Liv slumped back in her chair.

"What is it?" asked Angélique.

"It's like I live on a different planet. You have so many plants. Water falling from the sky. You eat cheese from cows. I've only read of these things. I mean, a cow. Think how much food a cow must eat. How much water it must drink, but you . . ."

Kasemchai placed his hand on her shoulder. "Liv. Do you need a moment?"

Liv composed herself. She pulled down on the front of her blouse and placed her hands on the table. "No. No, I'm fine."

She picked up the carafe and poured herself a drink. Kasemchai did the same.

"Well, shall we get started?" Angélique said.

Kasemchai's head snapped up. "Get started? What about Philippe?"

"What about him?"

"We're here to pitch our service to Philippe. I don't understand."

Angélique was solid as stone. "What's not to understand? You asked to set up a meeting with our family. I'm here now. Let's go."

Kasemchai pounded his fist on the table. Liv jumped. Angélique didn't even flinch.

He knew he couldn't do business if he lost his cool, but something didn't add up. How had he not seen this coming? "I specifically asked to set up a meeting with your father. What kind of game is this?"

"You asked to pitch your business to my family. I am obviously far more involved in my family business than you are. Now, do you wish to tell me about your business, or should we end this discussion right now?"

Kasemchai's eyes bulged, and the veins in his neck pulsed. Liv placed her hand on Kasemchai's shoulder and stood up. "It's not that Kasemchai doesn't care, Angélique. He talks about the families all the time."

It seemed like there was more to this thought, but she didn't continue. She waved her hand over her AugWatch, and the experience sphere in the center of the table pulsed green three times. "Did you know that Gouda is from the Netherlands? When I saw the cheese on the table, I thought, why should the place I was born deny me so much? I didn't burn the forests that covered the land in ash."

She turned and walked away from the table. Almost too quiet for the others to hear, she muttered, "Cheese. From a cow. How absurd."

She turned around again. "Noa, start my presentation."

The room around Liv darkened, and the scene behind her was underwater. The giant propeller turned slowly. As the image behind her crested the bow, Liv looked directly at Angélique and began. "Imagine the vastness of our oceans, where massive tanker ships traverse the waters, connecting nations and economies. These ships are colossal engineering marvels that have long symbolized the might of global trade. But they've also been symbols of something else—a challenge many believed was insurmountable."

Halfway through her presentation, Angélique began asking questions. The presentation didn't really finish so much as roll into a discussion of how many ships she wanted to convert.

"The Damen Shiprepair Rotterdam can berth several Coastal-class tankers simultaneously," Liv said.

"Good," Angélique said. "Then we start with five ships. If we can get those back in service in under six months, I'll commit to five more within the year."

The conversation was going well, and Kasemchai swelled with confidence. A few more agreements and Kasemchai knew they'd be on their way. "Not a problem. We'll be able to handle the full ten this year."

"What would you like us to do with the engines we remove?" asked Liv.

"Sell them. Use the proceeds to convert another ship or two. You're in a shipyard. I'm sure they can use those parts." Angélique selected a slice of cheese and placed it on a cracker. She took a bite and slowly chewed. "Now, I still have questions about battery-as-a-service."

"BaaS," Liv said.

"Yes, the BaaS. This is an ongoing fee. And how do I keep this fee reasonable?"

Kasemchai leaned in. "The service fee model means that you have a significantly lower maintenance cost. No fluids, no parts, a fraction of your crew."

"And we are covering the cost of retrofitting the first five ships," added Liv.

"Right," continued Kasemchai. "But the maintenance, charging, and swapping of the batteries will all be managed by Sol-vang. You won't own the batteries, but you won't own the cost of maintenance and replacement either."

Angélique dipped her head and with a raised brow said, "I get it, Kasemchai. My question is, how do we keep this as a reasonable fee?"

Liv pulled up a pie chart and pivoted it with her hand so that Angélique could see it. "We've, of course, modeled this out extensively. And we've come up with a model that you can benchmark. We can track the costs for fuel, removal of waste oil, bilge sludge, dumping fees, environmental fees, and engine maintenance fees."

As she listed each item, the wedge that represented each cost grew and shrank on the chart. "Our BaaS fee is lower than these combined costs, and we can track their annual increase and peg our rate to stay below that rate of increase."

"Okay. I can live with that, Liv. I have one more request, Kasemchai."

"Go ahead," Kasemchai answered in a guarded voice. He was realizing the questions to Liv were always tactical, but when she wanted to flex, she directed the questions to him.

"You couldn't convince your father, so you had to come to me."

"Well, that's not exactly true."

"It is absolutely true, Kasemchai." She calmly placed one hand on the table and then covered it with her other hand. "If your father were truly onboard, you would have led with that." She slightly tilted her head, the movement slow and intentional, like a knife twisting cruelly in his stomach. "I am your first major customer. This is a risk for both of us, and I expect to be rewarded for taking it."

Kasemchai shot a quick glance at Liv. "What were you thinking?" he asked.

"I want my services fees to be capped at five percent below whatever fee Preap agrees to."

Angélique smiled broadly like the cat that had just eaten the mouse. Not just eating the mouse but was actively chewing it right before him.

Kasemchai always won. It was easy—everyone wanted his money, and they were willing to bend over backward to get it. For the first time, he knew what that felt like. Worse, it would give the Labonne family a permanent advantage over his family.

He didn't look at Liv.

"Agreed."

Chapter Forty-Two

LUUK CAME to the office every day, but everyone knew he was doing absolutely nothing. After lunch, he scanned through the news on his AugWatch for an hour or two and was out the door by two. He'd walk across the Rijnhavenbrug, working up a good sweat and a foul mood, and wandered the streets of the Katendrecht district. Eventually, he would find his way into a bar; the darker, the better. Somewhere where they wouldn't know his name.

On this day, he found his way to a subterranean booch bar. He pounded two cayenne booches, each with a booster, then ordered a plate of bitterballen. The traditional fried Dutch snack was formerly made with ground meat. It was now typically a mixture of ants, beetles, and scorpions mixed with salt and garlic and then deep-fried.

When he asked for his check, the waitress pointed towards a table in the corner and said, "The table over there picked up your tab."

A woman and two men sat at the table. One of the men appeared to be much older than the other two, closer to Luuk's age. He noticed they all wore different clothes but somehow looked the same. Maybe it was the brown suspenders or the loose-fitting shirts.

"Dammit," he snorted. He got up and waved at the table. "Thanks." He hunched his shoulder and walked toward the door.

"That's a bitch about the Bundeswehr contract," the younger man said.

Luuk spun around. "What did you say?"

The young man waved Luuk over.

Luuk hesitated, then walked up and thumped his finger on the table. "You should be careful what you say about things that are not public."

The man smiled. "Please, have a seat. I think we have much in common."

Something seemed off about their appearance to Luuk—and then he realized none of them wore an AugWatch. This struck him as odd, as he hadn't seen anyone without an AugWatch in a long time. They didn't appear to be drifters; their clothes were simple yet clean and new.

The woman leaned in. "Do you find it odd that with billions of euros in investment and connections to the two water families, the only contract you've been able to land is with the military?"

"Wasn't the goal of Sol-vang to change the shipping world and eliminate one of the greatest sources of carbon dioxide?" the young man said.

Luuk pulled out a chair and sat down at the table. He grunted noncommittally. "It was. And to whom should I thank for picking up my tab?"

The older man grinned. "It is a gift from the Communist Revolutionaries."

The server leaned in and placed four cups on the table. Each person took a cup, and the older man lifted his booch. "We call ourselves the Rainmakers."

Luuk and the others all raised their cups and toasted.

"So you're all about collective ownership and class struggle, huh?" Luuk asked.

"There was a time when the Dutch were the bourgeoisie," the woman said. "It doesn't take a lot of imagination to figure out who the ruling class is now—the Khmer Empire."

"I don't follow you. The Khmer Empire? You're talking about Cambodia?"

"Angkor Wat was the largest temple in the world. Still is. Sophisticated water management made it possible with reservoirs, canals, and irrigation networks built to control seasonal flooding and provide water for agriculture—the very foundation of the temple's monumental structures. The temple's completion was also the empire's zenith, symbolizing its wealth, cultural sophistication, and devotion to Hinduism—reincarnation."

It was weird that they mentioned reincarnation, but Luuk skipped over that point for now. "You mean the empire ended after completing Angkor Wat?"

"Environmental changes brought drought. Crops failed. The empire relied heavily on trade and abandoned the upkeep of its water management systems." The hanging lights overhead cast dark shadows, and Luuk found himself staring at the old man's coarse grey hair on his chin as he spoke. "How could they just disappear? Water. What brought the Netherlands, Germany, China, and even the US to their knees? Water. Do you think it is a coincidence?"

"Well, yeah, it's a coincidence," Luuk said.

"No, they've been waiting for hundreds of years. Encouraging first-world countries to exploit the world's resources. Now they have all the water and the means to harness it, and we have nothing." The woman carefully pulled her long hair back from her face. "Kasemchai doesn't want Sol-vang to reduce pollution. He wants to rule the world."

Luuk tugged absentmindedly at the sleeve of his shirt. "You want me to believe that Kasemchai wants to rule the world? I know him. He's like a child. He doesn't know what he wants."

The old man squinted. "Maybe our story does sound fantastic. But outlandish stories have a kernel of truth. You can't deny that Kasemchai is using your batteries to work his way into the military."

The younger man studied the older man, then turned to Luuk. "Maybe he will build ships as well. Is that to reduce pollution or expand his control of water?"

"He doesn't build ships. We do." He placed one hand on the table and curled the fingers under his hand. "It's my company." *Or at least it used to be*, he thought.

The old man asked, "Do you know why we are called the Rainmakers, Luuk?"

"Of course. You want to bring rain to the Netherlands." He laughed. "Or maybe you're just great salespeople."

The older man laughed. "If we cared so much for money, we would be in a finer bar than this, I think. No, the Rainmaker is an ancient fable from Asia about a village suffering from a terrible drought. In desperation, they sought the help of a famous rainmaker who lived high in the mountains.

"He asked to be left alone in a hut and immersed himself in meditation and contemplation. On the third day, it began to rain. When the villagers asked how he did it, the rainmaker, with a humble smile, shared that he did not make the rain but simply aligned himself with the natural flow of the universe. When he found harmony within himself, the rains came naturally."

The man sat back and took a sip of his booch.

Luuk looked around the table. "Wait, you mean your group is named after a guy who did nothing?"

The old man took another sip of his booch and leaned forward, his hand still wrapped around the cup. "Think about it. The old rainmaker's ability to bring rain wasn't through some mystical ritual or incantation but rather through his own alignment with the natural order of things, or Tao, as it's known in Chinese philosophy. When the rainmaker arrived, he recognized that the village was no longer in Tao. As a result, he was no longer in Tao. It took him three days to bring himself into order again. Through his guidance and the collective effort of the villagers, they were able to restore balance to the natural world."

Luuk looked across the room at a fiddle-leaf fig plant in an ancient metal pot. The broad, pointed leaves were coated in dust except for the glossy streaks where people had reached out to confirm it was genuine plastic.

A look of irritation crept across the old man's face. He blinked, and the wrinkles vanished from his face. "When we are in tune with ourselves and nature, we can influence positive change in the world around us, even in the face of seemingly insurmountable challenges like drought. We are not radicals, Luuk. If you look inside yourself, I believe you will find a way to bring balance. You can influence positive change."

Luuk sat back in his chair and sighed. "It's a lot to think about, but I appreciate the booch. Next one's on me."

The younger man leaned in. "We meet here every Thursday, Luuk. For happy hour."

Chapter Forty-Three

KASEMCHAI EXITED the elevator on the thirty-third floor and walked towards Liv's office. He waved left and right at everyone on the floor and barged forward, finally stopping at a young man's desk.

"Hey, my name's Kasemchai."

The man looked trapped. He stammered, "Uh, yes, yes. I know that, Mr. Seng."

An amused smile played at the corners of his mouth. "What's your name then?"

"Aksel, sir."

"Great. Aksel, what do you do?"

Aksel stole a nervous look at his colleagues. "I'm, uh, I work with Liv on government contracts."

"Cool. Cool. Could you get me a cup of tea?"

"Yes, of course. Would you like hot tea? Jasmine? Milk tea?"

"Yep. Thanks. I'll be in Liv's office." Kasemchai strode up to her door and swung it open. "Hey Liv, how's it going?" he shouted.

Liv jumped in her seat. "What are you doing here? You didn't call to say you were on the way up?" She looked out the door and shouted, "Hendrik, could you get Kasemchai some tea?"

"I'm already on it, Liv." Kasemchai turned and shouted out the door, "No need for tea, Hendrik. I've got it taken care of."

Kasemchai sat down on the chair across from her desk. "Come on, Liv. I've been here so many times. You don't need to make a big deal whenever I come in. So what are we covering today?"

Liv got up from her desk to sit in the chair next to his. She placed the disc from her AugWatch on the coffee table. "Okay, we've contracted with Damen Shiprepair Rotterdam. Three of Angélique's ships are docked now. One is having the engine removed. The other two are in dry docks seven and eight. Those two are having batteries installed right now."

"Please . . . don't say that."

"Say what?"

"They are not Angélique's ships." His brows had drawn slightly downward, and his jaw tensed. "They are Philippe's. Angélique works for Philippe."

Liv scrutinized Kasemchai. "Sure, Philippe's ships."

"Thank you." Kasemchai relaxed again, as if the previous tension had never been there. "How long does the whole process take?"

"That is a really good question. I have a department analyzing ships now for this process. When we do this at scale, we will have to account for the type of ship, the age of the ship, the engine type, the dry-dock location—"

"Yes, but the three ships are in dry-dock now," Kasemchai interrupted her. "How long will they take?"

"Ah. Well, it takes about two months to remove the engine and clean the engine room completely. Then, two more months for the installation, which includes regular inspections. The final EU inspection could take up to an additional month."

Kasemchai leaned forward. "Five months? Are you kidding? That's awfully close to our deadline."

"Our pilot was on a much smaller vessel. These vessels are all over two hundred meters in length. But I'm aware of the timeline."

Aksel knocked on Liv's door. She motioned for him to come in. He walked directly over to Kasemchai and placed a cup of hot tea on the coffee table in front of him.

"Is this black tea?"

Aksel straightened quickly, like he'd been shocked. "Uh, no, sir. You weren't specific, so I thought you'd prefer jasmine."

"Jasmine? Why would you think that?"

"Uh, no reason, sir. I mean . . ." Aksel had turned bright red.

Kasemchai laughed. "I'm just kidding. You've got good taste, Aksel."

A bead of sweat formed on the side of Aksel's forehead and began to trickle down. He scurried out.

"Okay, what next, then?" Kasemchai asked.

Liv scowled, then looked down at the screen hovering above her AugWatch. "People have noticed what we're doing. We've received inquiries from six organizations, two of which have moved to the requirements stage."

"Really? Things are starting to get interesting."

"We've done quite a bit of analysis. Our current production capacity will not meet demand if we retrofit fifteen ships this year. We think it's reasonable to hit that target in year one. Based on our TAM, we need to rethink our manufacturing."

Kasemchai's brow knitted, and he squinted as he asked, "Tam?"

"Total Addressable Market. It means of all the ships in the world, how many could be retrofitted? We recognize that this idea won't catch on overnight, but the contract with Angél—I mean, Philippe, has changed the conversation."

She pushed the image out to the experience sphere with her hand so that Kasemchai could see what she was looking at. "Look at these articles." She pointed, and a journal headline came into view. "'Are ships finally catching up with electric vehicles?' *The European*." She swiped to the next article. "'What does the Labonne family know that the rest of the world doesn't know about electric ships?' *Al Jazeera*. 'Sol-vang is klaar om de wereld te veranderen.' *Quote Journal*."

Liv regarded Kasemchai's confused face. "Sorry, 'Sol-vang is ready to change the world.'"

"Okay, so what are you driving at, Liv?"

"It's time to build our own gigafactory."

"In China?"

"In Germany."

Kasemchai made an exaggerated sigh. "Figures. So, what's the catch?"

"We can start to set up retrofitting stations in Asia. This will accelerate the number of contracts we can close. We'll need to get a gigafactory online next year, but it will be expensive. This is risky because of the gap between potential business, real contracts, and real expenses."

Kasemchai sat back in his chair. Then he stood and walked up to the window. "This is risky, Liv." He paused and then spun around quickly. "And it's not like you. You're the sensible one. I'm the risk taker."

Liv smiled. "Nothing gets past you, Kasemchai." Liv stood up. She tugged down her blouse, then walked back to her desk and leaned against it. "We've just closed the deal on the Bundeswehr contract."

Kasemchai clapped his hands to his head as his eyes lit up with wide-eyed excitement. "*What?* How many?"

"Thirty armored ambulances and fifty command modules. We're also partnering with Mercedes-Benz to completely overhaul the chassis of forty-five Zetros off-road trucks."

Kasemchai leaped forward and hugged Liv. "You're a genius!" he shouted. "I can't believe it." He paced back and forth through her office. "More financing. Okay, let me think about that one. I'll need revenue projections. What else. Help with contracts in Asia?"

"We're already on it, Kasemchai . . ." Liz was trying her hardest to stay calm and in control, but the sound of pride in her voice was unmistakable. "However, we are going to need more financing for our factory. This is a big investment."

"So what's next?" He strode quickly to the darkened window and looked out over the city. "We need a drink."

Liv smiled her impish smile. "That's why I invited you here this week, Kasemchai."

"Why?"

Liv swung her arms out wide, insinuating the date should be obvious to everyone. "Friday is Koningsdag. The King's Day."

"Okay. So what's King's Day? We light some bonfires?"

She wagged her finger at him. "Kasemchai, we don't light bonfires. Did you forget the planet is shit?"

"I'm sorry. Tell me about this little holiday."

"Oh, it's just a little holiday that began in 1885. There's music and flea markets, and everyone wears lots of orange. In Amsterdam, it's a tradition to organize a party on a boat and cruise down the canals."

"This sounds awesome." Kasemchai cocked his head to the side. "But Liv, the canals lost their water years ago."

She grinned broadly. "Not to worry. We have a plan."

KASEMCHAI FOLLOWED Liv's directions and walked along Rozengract towards Prinsengracht, the Prince's canal. He could hear a steady drumbeat, and the crowd became denser as he neared the canal. Everyone was wearing orange—hats, shirts, pigtails, and pants. He passed six women, all wearing matching blue pants and orange shirts.

When he reached the canal, he saw that the banks were filled with people dressed in orange. Nearly everyone had a cup in hand, and they rhythmically swayed with the drumbeat.

Liv came up behind him and grabbed his shoulder.

"You made it!" she shouted. She hugged him and handed him a plastic lei of orange flowers, which he placed around his neck.

"Are we going down there?"

"Yes, of course."

Kasemchai looked down into the dry canal and saw it was filled with bicycles. Many of them had a platform with four fat tires and six or eight seats along two sides. The people leaned forward over a table that ran down the center. It was, he realized, a bar to hold their drinks.

"Are we going on one of those?" asked Kasemchai.

"Of course," Liv giggled. "We used to ride boats down the canal. There's no more water in the canal, but we love our bikes. So we improvised."

They climbed down the ladder and up to a barcycle, where he saw Johannes, Chantal, Helena, and Hendrik. The bike had four tall posts on each corner with horizontal boards running corner to corner. They were painted with tulips, flowers, and booch cups, and strings of lights draped down from each corner.

"I thought Luuk would be here?" asked Kasemchai. "He loves his booch."

"Back in the day, he did. Now, he leaves the office every day early to sulk. Who knows what he's up to?"

Liv took a seat on the barcycle and held her hand out, inviting Kasemchai to sit in the empty seat next to her. Hendrik passed a booch to them, and everyone raised their glasses.

They shouted in unison, "Proost!"

The barcycle lurched forward, and they pedaled their way into the parade of orange headed down the canal.

Hours later, Kasemchai and Liv sat across from each other. The sun began to rise, and the crowd became eerily quiet. There was a shout here or there, but most people put down their glasses for one moment to watch the blazing orange spread across the skyline.

Kasemchai tipped his cup toward Liv's and gave it a little tap. "Thank you."

She smiled back at him. "Of course."

Kasemchai leaned close to talk into Liv's ear. "Hey, I have an idea. Want to come to Cambodia with me for the Water Festival?"

Liv placed her hand on Kasemchai's thigh and turned to face him. "That sounds beautiful. What is it?"

"We go down to the river and we pray for peace to Mae Kongkea. She's a goddess who we believe brings the water for people and animals."

"Do you believe she brought the rain to Cambodia?"

"I suppose so. But the festival has changed because of it."

"In what way?"

"The festival used to celebrate the end of the monsoon season. It was a way of knowing that we were now safe. Now, we celebrate the monsoon that never ends. It's a tragedy and a treasure."

Liv gave his thigh a gentle squeeze of reassurance. "In what way is it a tragedy?"

"Well, of course, you know many people died around the world when the rain changed. In Cambodia, the main festival used to be celebrated in Sisowath Quay. That city is now in the center of the Rain Wall."

"What happened to the people there?"

"They're gone. No one survived behind the wall. We were lucky to be on the outside."

"I'm sorry." She looked down and leaned in closer so that he could hear her speak. "I never really thought about how the people in Africa and Asia suffered. Too many people around us were dying."

Kasemchai stared down at the dusty ground. He placed his hand on hers, then he lifted his head, grinned broadly and asked, "What do I need to do to get a drink here?"

Chapter Forty-Four

THE BUNKER BARGES were in operation at all hours, every day of the year. The crew worked twelve-hour shifts, generally no more than fourteen days a month, to ensure they had enough rest between shifts.

The crew began the day with a detailed handover between officers and non-licensed members. Makoa was still one of the non-licensed members. Their positions had names like Able Seaman or Ordinary Seaman.

One year into the job, Makoa recognized that his life had become exceptionally ordinary. He enjoyed the extra time with his family, but on the barge, his role was moving heavy equipment, cleaning, and waiting. There was a lot of standing around and waiting.

They had just finished bunkering a tall cargo ship. As they pulled away, Makoa found himself in his favorite spot: the aft railing. It was his sanctuary, a place of solitude and contemplation where he could gaze across the bay. He was studying one of the medium-sized tankers across the bay when he realized Ben had come up beside him.

Makoa gestured across the bay and yelled to be heard over the droning engine noise. "Hey! I noticed we don't bunker some of the water tankers, like that one with the Labonne crest. How do they fuel up?"

Ben looked down at the deck and spat. "They don't. They ripped out the engine and replaced it with batteries."

Makoa's brow furrowed as he processed this. "No kidding?" He laughed. "Do they have to bring out big cables like an EV?"

Ben nodded his head. "Yep. Sometimes, cranes remove the old batteries and replace them with new ones. I don't understand what's happening, but the entire process takes just over an hour."

Makoa thought about that. "So they don't need a bunker barge?"

"Nope. You picked the wrong time to join this industry, Makoa. Just a matter of time before all ships are electric. Who needs a tanker-man, then?"

Makoa stared out at the ship across the bay. He thought about the money he had spent on training to be a tankerman. And the money he had borrowed. The person at the school said it was a solid industry with great benefits.

He thought back to the night in the bar when he complained about Kasemchai. All he had to do was keep his mouth shut, and the winery wouldn't have burned down. If only he hadn't been so stupid or selfish. He could almost imagine that night going differently.

The engine shuddered beneath his feet, and he recognized his past was as dark as his future.

Chapter Forty-Five

THE CAR CONTAINING Liv and Kasemchai departed from China Resources Headquarters in Shenzhen. The towering building was made of metal and glass, wider in the middle and pointed at the top like a bamboo shoot sprouting from the ground.

Kasemchai leaned over the armrest. "That went better than expected. Can you ship the batteries in time for the launch?"

"You thought I would make that commitment without meaning it?"

"Well . . ." he dragged out. "Sometimes I make commitments based more on what people want than what is possible."

Liv dropped her chin and smirked. "I've noticed. But that is not the case. We're ready to begin shipping batteries to Shenzhen immediately. But to sustain this, we need to close the deal on the EC."

"EC?"

"The energy center. Where we store and charge the batteries. Kasemchai, you need to pay attention to these details."

Kasemchai grasped her wrist and gave it a gentle shake. "That's why I believe in you, Liv so that I can focus on the relationships. Anyhow, what do you want to do before dinner? Fancy a cocktail?"

"I have plans for us," Liv said.

"For what?"

"When we were upstairs, you looked down on what looked like giant river stones."

Kasemchai scrunched his eyes, recalling the view from high above. "Yeah."

"They're actually several buildings by Ma Yansong, one of which is a museum."

"Should I know him?"

"I don't think you would know him. But do you remember when I took you to the Rotterdam Center for Dance?"

"Sure, with the winding staircase. That was cool."

"Ma Yansong designed both of these locations. So we are visiting the museum in Shenzhen Bay Culture Park."

Kasemchai nodded in affirmation. It was neither disdain nor joy, just a confirmation that he heard her speak. His head slowly rolled to the side as he considered how to make this day a little more interesting.

The streets were oddly quiet on the drive. This was a city built for twelve million people, half of whom died during the drought. They left the shadow of the tall skyscrapers and drove over a short bridge and into the park.

Kasemchai stepped out of the car and peered up at the smooth grey shapes. He thought they did look like giant river stones. On his right was a giant mound. On his left, the largest stone pointed straight up, with a long, thin stone shape emerging from the ground and leaning against it.

It was disorienting. They looked exactly like stones, but as they approached, the smooth sides loomed over their head. The meeting left him thinking how much he and Liv had accomplished, but when he beheld the enormous stones, it struck him how little they had accomplished. Every step forward was long and painful, yet only a few ships had been converted. The vast majority were still relying on bunker fuel.

A docent greeted them and led them inside. During the tour, they encountered an atrium featuring a reflecting pool at the center of the building. Glass panels stretched from floor to ceiling. He touched one of the panels and felt it was cool to the touch, shielding him from the outside. An enormous river stone was visible at the center, seemingly holding the weight of the building above it in a stubborn and precarious way.

Kasemchai looked down at the curved stone benches surrounding the glass and then back at the center river stone. It was massive and heavy, but it also seemed like it could be pushed over if he touched it in just the right spot.

He turned to Liv and the docent. "Why don't you continue without me? I need to make a call."

"I hope this is not an excuse to get out of our tour," Liv said sternly.

"No, no, no. It's business. I'll catch up shortly."

As Liv and the docent walked away, Kasemchai glanced down at his AugWatch. "Chanthavy, call Dad."

Preap's avatar appeared, looked down at the bench, and sat beside him.

"Hi, Father. How are you?"

His father didn't hesitate but moved right into business, as if he had called Kasemchai and not the other way around. "I'm fine. Your mother said to remind you that you need to be here two weeks before Bon Om Touk. So many things to do."

"Don't worry. I'll be there. I'm in Shenzhen right now."

Preap's avatar froze for a long moment. "What are you doing there?"

"I'm in a museum."

"Really? Are you buying it and tearing it down?"

Kasemchai placed his hand dramatically on his chest, feigning insult as he replied, "No, I'm looking at art. Liv brought me here."

"Liv? From Sol-vang? She brought you to a museum?"

"That's not so strange, is it?"

"My son is in a museum for no reason but to be a patron of the arts. Yes, that's unusual. This Liv must be an extraordinary person." Preap spoke so loudly Kasemchai feared a couple walking around the monolithic stones might hear him, before remembering the audio was in his ears only.

"She is, but not for dragging me to a museum. We just met with the Port of Shenzhen and the finance ministry. We've entered an agreement to run our BaaS model through the port. Three companies, including Philippe's, are already ready to send electric ships through here."

"Ah . . ." There was a long pause before he continued. "So, that's what this is about."

"It is. You said you'd commit after Philippe retrofitted ten of his ships. We've exceeded that number already." Kasemchai waited.

"I'm surprised Philippe accepted your offer. As much as he complains about the cost of oil, I didn't expect him to try something so innovative."

Kasemchai grinned.

"What's so funny?"

"I didn't meet with Philippe."

Anger washed over Preap's avatar's face. "What do you mean? He didn't have the respect to meet with my son? Who did he send in his place?"

"Angélique."

His father's avatar surged to its feet. "*Angélique*?"

It caught Kasemchai off guard. He looked around quickly, realizing again that no one could hear his father but himself. He let out a little quiet chortle. "Why does that surprise you?"

"Ten ships is a decision not only about the cost of engines. Ships in dry docks don't ship water. I can't believe he'd let his daughter make this decision." He paced back and forth. "Of course, it's a lesson. He already made the decision."

"Why does that matter? She's a shrewd negotiator."

"Because this isn't the type of decision he'd make."

Kasemchai stood up and faced his father. "Perhaps you don't know him as well as you thought." He looked at the ground. He didn't mean to be so argumentative. "Or he's changed." Kasemchai stepped back and sat down again.

His father's avatar sat down as well. "You may be right. Perhaps he has changed." He looked up toward the sky. "How many ships exactly?"

"Twelve."

"All right, I'll commit to fifteen ships."

Kasemchai grinned despite his effort to stay cool.

Preap's avatar stared deeply into Kasemchai's eyes. "What did she tell you to do with the engines?"

"She told us to sell them. We committed to covering the cost of retrofitting the first five, which we'll do for you as well. She probably wants to use the revenue to offset some of the lost water revenue."

Preap's avatar smiled. "She's a fool. I want you to store all my engines. Believe me, no one wants to end oil more than I do. But ships are a thirty-year investment. I need contingency plans."

Kasemchai stood and placed his hands together high in front of his face. He bowed slowly. "Thank you, Father."

His father's avatar bowed back. "Don't thank me yet, Kasemchai. Change is coming. Change makes me nervous."

Chapter Forty-Six

THE GIRLS SHARED A ROOM, the two beds on opposite walls. When Makoa came in to tuck them into bed, Kailani was already under the covers. Apikalia sat on the floor, playing with her stuffed frog.

Makoa pursed his lips. His voice was exasperated when he said, "Please, Apikalia. I told you to get into bed."

Without looking up, she responded, "It's not even late. And Mr. Frog still needs to brush his teeth."

"Frogs don't have teeth. And you need to get into bed." He smiled, bent over and gently picked her up. With his left hand, he pulled back the covers and placed her on the bed. "Come on now, fluff your pillow, and I'll tuck you in."

Apikalia reached over to her night table and drank from her bottle of water. "You know, Daddy . . ."

Makoa sighed. "Just lie down. Then you can ask me a question."

Apikalia fluffed up her pillow but didn't lie down. She tried to ask her question again, and Makoa again, more forcefully, told her to lie down first. Finally, she relented and lay down on her pillow. Makoa pulled the covers over her.

"You know what, Daddy?"

"What is it?"

"Are we going to go on vacation this summer?"

Kailani sat up in bed, and the two girls chanted, "Vacation, vacation, vacation!"

"Uhh, don't remind me." Then he smiled, pulled the cover back over Apikalia and kissed her on the forehead. "Good night, kaikamahine."

She responded, "Good night, makua kāne."

He sat on Kailani's bed and kissed her on the forehead. "Good night, kaikamahine."

She smiled back and said, "Good night, makua kāne."

He got up and went to the door, slid his hand over the shiny white panel beside the door, and the lights quickly dimmed and went out. "Aloha pō."

"Aloha pō," the girls sang back.

Makoa headed down the hall towards the living room. He stopped in the kitchen to grab a sachet of kombucha. The sachet, made of seaweed, was both a container and a snack. It was an acquired taste.

Anela was sitting on the couch. She was watching a virtumersion on the experience sphere while folding laundry on the coffee table in front of her. He slumped onto the couch, opened his booch, and sighed loudly.

"What's on your mind, babe?" Anela asked.

"The girls deserve a vacation. *We* deserve a vacation."

Anela continued to fold laundry and without looking up asked, "You want to go on vacation this year?"

"Sure I do. But we don't have the money. You know that."

She paused her folding to think. "We could stay with family. Just so the kids feel like there's a change."

Makoa sighed deeply. "I don't know."

"Don't worry, Makoa. The girls know you have a new job. It's going to take time to get on our feet."

"Yeah," he said, not sounding convinced. He stared at the wall and took a long sip of his kombucha.

Anela continued to fold the laundry. "How much clothing do you guys have? I feel like I'm folding laundry every day."

Makoa grinned. He placed his hand on a shirt as if he were about to fold it. But his hand just lay there. Finally, he took another drink from his kombucha, and his smile soured. "You know what they're calling us? Luddites."

"What's a Luddite?" Anela asked.

"Someone afraid of new technology. There was a bunch of guys in old England that trashed all the new equipment in their factory. They called them Luddites for some reason."

Anela looked up quickly. "We don't need any more fires, Makoa."

Makoa pressed down firmly on the couch cushion with his free hand. "I told you. I wasn't there when the winery caught fire. Not that it matters. I still lost my job. And I'm not a Luddite. I don't care if they run ships on batteries as long as there's still ships for us to fill with crude."

Anela dropped her laundry in her lap. "Are you going to lose your job?"

"No, no. There's nothing to worry about. Bunker boats have been around since coal. As long as there's engines, I'll be in business." He stared out the living room window at the lights twinkling from the balconies across the courtyard. "Besides, Ben is going to retire soon. That means there's a new position opening for me."

"Good. And we need more water."

"Again? I just bought water."

"You bought water on Saturday. It's Wednesday."

Makoa let out a sigh as he drained the last of his kombucha and headed to the kitchen. He stood over the trash can, tore off strips of the sachet, and ate them. He needed a little energy before he made the trip down to the water vending machines on the ground floor.

From the living room, Anela's voice called out, "Pick up twenty-five gallons, Makoa. Then you won't have to get water again until Friday or Saturday."

Makoa grumbled to himself, "Sure, maybe I'll pick up some caviar or ice cream while I'm down there. It's only money."

Chapter Forty-Seven

LUUK ENTERED the bar and walked straight toward the back of the room. He waved to the bartender and put up a finger for his usual, a cayenne booch with a booster. The Rainmakers were seated around their table.

"Any good news in the shipping world?" asked the older man.

Luuk pulled out a chair and sat down. "Yeah. We're retrofitting ten more Seng family ships."

"What does that mean?" asked the girl.

"We're replacing the engines with batteries."

"That's good," the younger man said. "Isn't it?"

"That is good," the older man said. "Sol-vang is making a difference. Less carbon spewing over the ocean."

"Yeah," Luuk said. "I suppose so." He took a long sip from his cup. "But the Seng family wants us to save all of their engines. If Papa Seng isn't happy, he can just put the engines back in the ships."

"That's bad," the woman said. "He still controls the natural world. A few ships run on batteries, but they still have thousands of ships that run on oil."

Luuk grunted, "And the Labonne family has thousands more. If we can prove they're more efficient than oil, we can remove tens of thousands of cubic tons of CO_2 in a year. But if they change their mind and return the engines, we are obligated to put them all back into production."

The young man sat up straight. "Then we have to make sure they never do. Can we do that?"

Luuk laughed. "That's a little harder to do than you think. Sol-vang is a business. We can't just refuse to put them back."

The older man grinned mischievously. "Sol-vang can't."

Luuk studied his cup. "That's right. Sol-vang can't." He swirled his cup around in a little circle and then froze. He lifted his head, and his usual sour demeanor vanished, replaced with a triumphant smile. "And the engines stored here are insignificant. But if production for the Seng family ramps up, there will be thousands of engines in storage. Probably somewhere in China."

The young man sat back in his chair and smiled at Luuk. "I think you've found your path, Luuk."

Luuk slowly nodded in the affirmative. "Yes, I have."

"So, how do you keep the engines from being returned to the ships?" asked the woman. "Do we pour some sort of gunk into the engine?"

Luuk shook his head. "No. These are huge machines, thirty or forty feet tall. Incredibly complex, with a main engine, a fuel system, a cooling system, and an exhaust system. The pistons alone can be over ten feet tall."

The woman's face fell. "Oh, this is going to be complicated."

The older man raised his hand triumphantly. "Saltwater? Could we put saltwater inside all the open parts to corrode them?"

Luuk's face brightened. "You're on the right track, my friend. Corrosion is the answer, but it's not that easy. All the open ports would be sealed. The engines are made of cast iron or steel and coated with a corrosion-resistant coating. They're made to work on the sea."

The young man laughed. "So, do you have an answer or not, Luuk? The suspense is killing me."

Luuk turned to him and said, "I do. It's a colorless gas that smells like rotten eggs. And it's not that hard to get. In fact, in a wonderfully ironic twist, it's a by-product of petroleum: hydrogen sulfide."

"It's too perfect." The older man squeezed the younger man's shoulder and turned back to Luuk. "Angkor Wat succumbed to the jungle, and now a natural gas could destroy their engines centuries later."

"That's right," Luuk said. "All we need to do is release the gas near the engine parts. It penetrates the protective coating and reacts with the moisture in the air to create sulfuric acid."

The older man raised his glass. "It's like the Rainmakers are destroying the engines with acid rain."

Luuk grinned. He raised his cup and held it aloft. "To acid rain, comrades!"

The others raised their cups and slammed them together. "To acid rain!"

Chapter Forty-Eight

KASEMCHAI THRUST off the pool wall, shooting straight forward under the water, sending ripples to the far edges. The water bounced off the gold-tiled lip and made a slight sucking sound as it fell back down on itself. He swam underwater, holding his breath until he could feel his entire body pulse with each heartbeat.

At the far end, he touched the edge and turned to gasp for air. He stared up at the mosaic ceiling. The swaths of gold finish pulsed with the water that he disturbed below. He closed his eyes and felt the warm sun stream directly overhead through the large circular window.

His wrist hummed repeatedly until he recognized someone was trying to reach him. He glanced down and said, "Chanthavy, answer the call."

Angélique's avatar stood over the pool, looking down at him. "Is this a good time to talk?"

"Yeah," he panted out. "Just doing a few laps."

A grin passed her avatar's lips. "In that case, I had better get into something more appropriate."

Angélique's avatar shimmered, and her clothes transformed into a bikini.

Kasemchai's gaze followed the strings that radiated out and over her hip. "Interesting choice for a married woman."

"We mustn't dull the senses, Kasemchai. Besides, my clothes were dirty." Angélique's avatar grinned mischievously, walked up to the ledge, and sat down. It dangled its legs in the water and kicked playfully.

Kasemchai felt his body rise in temperature as he remembered the day when she led him around her parents' house on the Ivory Coast. They found themselves on a lower floor, a darkened laundry room. She had jumped up onto a warm dryer that was tumbling clothes. It spun faster and faster, and the dryer began to shake as she stretched her legs out, beckoning him. Kasemchai had stepped up and grabbed her hips. That was their first time.

"I wanted to tell you that you have some competition."

Kasemchai's head turned sharply. "What do you mean?"

"The UN Environment Programme has launched the Emerald Initiative. It challenged oil-producing nations to create cleaner fuels. The Saudis have announced they are producing a less viscous fuel that they say is cleaner and doesn't need such a long warming time to be ready for burning."

He didn't like how she reported this news to him. She wasn't sharing it with him to be helpful; she was teasing him. He was certain something bad was coming next.

Kasemchai leaned forward. "So what? Less viscous fuel. What does that have to do with batteries?"

Angélique's avatar had an impish smile. "That's not the best part. This fuel will cost significantly less."

The best part, he repeated to himself. *She's toying with me, but why?*

Kasemchai looked down. He pressed his hand through the water, watching the water flow over his fingers. "Like I said, who cares about less viscous oil? It's still oil."

Angélique laughed. "No one cares about thinner oil. It's the price. The Saudis are selling thinner crude at a discount so that they can use the incentives from the Emerald Initiative to launch a new battery factory that they say is more advanced than the Sol-vang batteries."

"It's hype. They don't know anything about batteries." He knew that there was a punchline. *Why doesn't she get to it?* "And?"

"And what?"

"And what does this have to do with me? With Sol-vang."

He pulled himself out of the pool and walked towards the lounge chairs. With a plush towel, he rubbed his hair vigorously, wrapped the towel around his shoulders, and dropped into the chair. He refused to feed into the joke she so obviously wanted to spring on him.

Angélique's avatar walked over and stood in front of him. "Don't you get it? They've taken away the financial incentive. Why would I convert any more ships when they're practically giving oil away?"

"Shit! I hate environmentalists." He slammed the cushion as he stood up and began to pace. "They completely miss the point. People are driven by money. This stupid idea to make oil cleaner is going to sabotage my work, which"—he wagged his finger at her—"is truly carbon-free."

Angélique laughed so hard her avatar doubled over, clutching its stomach. It held its hand over its mouth while trying to find its way to the lounge chair to sit down.

He gaped at her in astonishment. "Now what's so funny?"

She took a moment to rein in the laughter. "You think the environmentalists are behind this? It's our fathers."

"What?"

"Well, you always said you wanted to negotiate with the Saudis. It turns out you've been part of the negotiation all along. Only no one told you your part."

Kasemchai balled his fists, extended all his fingers, and then balled his fists again. He wondered how he could have been the fool. Was it possible his father was playing him this whole time? "You want me to believe that your father and my father schemed to get me to invest in a battery company . . ."

He spun around. The pool's surface danced across the glittering mosaics overhead, where an intricate mural of the nāga—a mythical seven-headed snake—twisted among clouds of jade and lapis lazuli. He could almost feel the steam rising from the stiff helix of his ear. "Did you send me to Ishigaki Jima?" he demanded.

"No, Kasemchai. I did not send you to Ishigaki Jima as an elaborate ruse for you to invest in a battery company." Angélique's avatar crossed its hands behind its head. "But they needed to put pressure on the Saudis, and converting ships to battery power was exactly what they needed."

Kasemchai angrily cut through the air with his right hand, his index finger jutted out like the barrel of a gun. "That doesn't make any sense. They know that the oil is going to run out eventually. What kind of game are they playing?"

Angélique's avatar shimmered, and then it was wearing a white robe wrapped tightly around its body. "Kasemchai, there's only one team in this game: our fathers. And the game is to win all the power and money."

He stormed up to Angélique's avatar and stood over it, his fists clenched with restrained fury. "What about the future? They can't just burn oil until it runs out."

"Power and money, Kasemchai. Don't ruin our legacy."

Chapter Forty-Nine

KASEMCHAI RODE the elevator up to Liv's floor. He hadn't been to the office since they celebrated King's Day the previous year. The doors opened, and he purposefully strode out of the elevator, bumping right into Hendrik. Both men stepped back, surprised.

"Excuse me, Mr. Seng," Hendrik said quickly.

"No problem, Hendrik. Why are you standing so close to the door?" Kasemchai asked as he continued towards Liv's office.

"Liv asked me to make sure I didn't miss you when you arrived, sir," Hendrik replied.

Kasemchai stopped and grinned at Hendrik. "Did she? So, she wants you to ensure I don't wander off, then?"

"Oh no, sir. She just wanted to make sure you were properly greeted." Hendrik stood back to collect himself.

Kasemchai continued walking to Liv's desk. He surveyed the floor until his gaze stopped at the desk where he had stopped on his last visit. "Aksel," he called out.

Aksel whipped his head around and jumped to his feet. "Yes, sir, Mr. Seng."

"Are you still in contracts?"

"Yes, sir."

"Great! It sounds like you're ready for a promotion. I'll talk to Liv about this. Could you do me a favor?"

"Of course, sir." A sly grin spread across his face. "Would you like hot tea or cold tea?"

Kasemchai winked. "Yep." He turned and strode towards Liv's office.

Hendrik followed, trailing behind him with his arm outstretched. "Right this way, sir."

Kasemchai laughed. "That's a lot of 'sirs', Hendrik. Aren't we on a first-name basis by now?"

"Of course, Mr. Kasemchai. I mean, Kasemchai." He then mouthed, "Sir," just to be safe.

Liv was waiting for him at the other end of the room, standing by her open door. She greeted Kasemchai, "Are you thirsty?"

"No, I'm not thirsty," he replied.

They embraced warmly, then stepped back from each other. Liv placed her hand on Kasemchai's shoulder and drew him back in closer.

"It's so good to see you, Kasemchai. And I see you ignored all my efforts to greet you properly." She turned to Hendrik. "I'm sorry. He can be hard to manage."

Hendrik rubbed his hands together nervously. "I tried my best, Liv."

"He certainly did," Kasemchai said. "He nearly knocked me over as I entered."

"Good, good! Just as I instructed," Liv laughed. "Please, come on in, Kasemchai."

Inside her office, Johannes stepped forward and extended his hand. "Good to see you again, Mr. Seng."

"Of course. You too, Johannes."

Liv called over to Hendrik, "Could you get Kasemchai a drink, Hendrik?"

"No need. Aksel is already on it," Kasemchai said.

Liv frowned. "Could you at least try to let people work on their assigned tasks? Aksel is working on a very important project. In fact, that's the one I want to review with you today."

"Excellent! I was going to mention that I think Aksel deserves a promotion. He's a great guy."

"Thank you for that unsolicited suggestion, Kasemchai. Well then, let's get started." She turned toward Johannes. "Would you pull up our presentation?"

Kasemchai took a chair while Johannes toggled the colorful bubbles floating above his AugWatch. The experience sphere cast his presentation in front of the wall for everyone to see. It showed a map of South America, with the Brazilian border outlined. "I believe you're aware that we are now actively building electric armored vehicles in Germany, the United Kingdom, Italy, and Korea."

The map zoomed out to be a globe that slowly rotated, with each country clearly outlined. "Next month, we're finalizing an extraordinary deal in Brazil. We're partnering with the Chinese and Brazilian governments to build military vehicles in Recife, Brazil." Johannes double-tapped his fingers, and six military vehicles appeared, including heavy tactical trucks, light cargo trucks, an air defense radar truck, and an armored patrol boat. "This joint agreement will produce twenty percent of its vehicles for Brazil. The remaining eighty percent will be exported to countries including France, South Africa, Chile, and the Netherlands."

Kasemchai pursed his lips and nodded in affirmation. "This is very creative. How on earth did you develop this partnership?"

"Good question. The Chinese government wanted to double production in Shenzhen. We were working on several shipping contracts when I brought up the work we were doing developing military vehicles. They were hesitant to work on any of their own military vehicles, but our good friend Aksel did a little research and discovered this initiative the government had with Brazil."

Kasemchai grinned and stared out the glass wall of Liv's office towards where Aksel sat. "See, I told you he needs a promotion."

Kasemchai appreciated Liv's scowl. It meant that she was no longer intimidated by him and that he could treat her like an equal. He also enjoyed bringing a little levity to any conversation, which kept people on their toes.

"Sorry, as you were saying, you're doubling production in China," Kasemchai said. "That's amazing."

"That's not the big news. This agreement opened up another

opportunity. We're opening five ECs in Brazilian ports." Liv swiped her hand in the air and the map returned. She pointed to glowing dots that appeared on the map. "Two deepwater and three smaller ports. This is a major expansion of our infrastructure across the eastern side of South America. As a result, we should be able to escalate our conversion plan with Mr. Labonne."

Kasemchai lifted his hand and gazed at it, tracing the lines in his palm with the tip of his finger. He sighed, then asked, "Have you heard of the Emerald Initiative?"

Liv stared at Johannes, who answered for her, "Yes, it's a UN green initiative, I believe, focussed on cleaner fuels."

Kasemchai looked at Johannes and nodded his head in affirmation. "Well, it's a little complicated, but that initiative will subsidize the cost of a new oil the Saudis are producing. My father and the Labonnes probably won't be starting any new battery contracts in the near future."

Liv stared out the window at the dark sky. She turned back to Kasemchai. "Did your father and Mr. Labonne tell you that they were putting a halt to new contracts?"

Kasemchai glowered. She had done it again. Kasemchai always acted with his heart, which meant he was easily worked up. Liv always processed things in a very analytical way. Was he overreacting?

"No, they didn't," Kasemchai answered. "But I spoke to Angélique, and she told me as much."

Liv crossed her arms in front of her. "We shall see then. We still have many military contracts and many other shipping companies to work with. The Saudis can't subsidize all the oil in the world."

Kasemchai sat back and stared at the ceiling. He blew out slowly through his pursed lips. Then, he stood abruptly and grinned. "Liv, I think it's time we go back to the scene of the crime."

She looked at him like he'd lost his mind. "Huh?"

"I want to go diving in Ishigaki Jima again. Visit the captain and see how he's doing. We've made great progress. I think it's time to recharge, well, spiritually." He glanced over at Johannes and nodded. "And we've had a couple of challenges. But nothing we can't overcome. We need to step back and see where this thing is going."

Liv grinned. "Right now?"

He loved it when his spontaneous ideas bloomed into a plan.

"Absolutely not."

His mouth fell open. "What do you mean?"

She blinked, then opened her eyes wide, waving her hand toward the office. "Because there are way too many projects going on. It would be irresponsible of me to step out and take a little trip to a tropical island."

Chapter Fifty

LIV CHECKED her AugWatch in the backseat, then looked up with a scowl as the car halted. "I can't believe you."

Kasemchai regarded her with mock surprise. "Look, the best ideas don't come from work. They come from when your brain rests. And your great big brain needs a great big rest."

He leaned forward and opened the door. The sun streamed into the car, and Liv reflexively threw her hand up to protect her face.

"Wait!" she yelled. "It's still daytime." Then she chuckled. "Sorry, I forgot how much cooler it is in this part of the world."

She leaned forward and followed Kasemchai out of the car. She looked around, bewildered. "I thought we were going straight to your yacht?"

"One pit stop first."

He pointed toward a brightly painted building. They walked down the narrow street to the tiny Baga Jima dive shop.

"They've made improvements," Liv observed.

"What do you mean?"

"It's still small, but it looks much nicer than I remember. There's a new awning, and it's been painted."

Before they could knock, the captain slid back the door and, with great gusto, bounded outside. He stopped short and, with his hands on his knees, gave a quick bow.

Liv and Kasemchai bowed back. Kasemchai clasped his hands in front of his face as he often did.

"Are you thirsty?" asked the captain.

"I'm not thirsty," responded Liv and Kasemchai.

The captain threw his hands up in the air, and Japanese flew from his mouth.

"Woah, woah, woah," Kasemchai said. He put his hand up to stop the captain.

Liv and Kasemchai asked their AugWatches to translate for them. The captain fumbled over his AugWatch to get it to translate as well.

Once his translator was active, Kasemchai said, "That's better. Now, what were you saying?"

"I wanted to say, welcome back to Ishigaki Jima! You are my most famous dive guests." He pointed excitedly at his dive shop. "Everyone wants to go on the dive boat where the famous Kasemchai and Liv first met." He leaned in conspiratorially. "I have a newer boat, but no one needs to know that."

Kasemchai peered beyond the captain and saw Haruka rummaging around in a tub of dive gear. "Hey, Haruka! Are you going to join us for dinner tomorrow?"

The woman stood up and smiled. Kasemchai realized immediately that it wasn't Haruka.

"I'm sorry, I thought you were Haruka," he called out again.

The captain laughed. "No, that's Yuki. Haruka hasn't worked here for several years. Not since she was married."

Yuki came out to greet them. "Welcome back to Ishigaki Island. The captain often talks about his famous dive with you two."

Liv shrugged. "I'm glad he still remembers us."

"Of course, Ms. Liv. And he said you are a very good singer, Mr. Seng."

Now Kasemchai threw his head back to laugh. "I think the captain had too much to drink if he thought I was a good singer."

"Would you like to come inside?" asked Yuki.

"That's okay," she said with an easy smile, tucking an errant hair behind her ear. "We were just heading down to the beach and wanted to stop by. Do you know if Haruka will be able to join us tomorrow?"

Yuki gave a slight bow. "I'll check. Her son has a cold, but I know she was planning on it."

Kasemchai looked at Liv in surprise. Her eyes told him she was just as surprised.

"Is that possible?" he asked Liv.

"I suppose so. It's been more than five years since we were here."

The captain patted Kasemchai on the shoulder. "Don't worry, we know you two have been very busy."

They said their goodbyes and headed to Kasemchai's yacht.

THE NEXT EVENING, Kasemchai and Liv hopped into the car and headed back to the dive shop. When they arrived, Haruka and the captain were standing out front.

Liv darted out of the car. "So good to see you, Haruka!"

The girls hugged. Kasemchai climbed out next and gave Haruka a big hug.

The captain grinned at everyone and then looked over Kasemchai's car. He gave a long whistle. "My, my, my. I don't think I've ever seen a car like this on the island."

Kasemchai reached out and patted one of the large knobby tires. "I keep this in my yacht. You never know when you need to get around on sand dunes."

"Sure, I get it," the captain said, but he didn't look like he got it.

Liv beamed at Kasemchai, then looked back at the duo. "We owe so much to you two. Everything Kasemchai and I accomplished wouldn't have been possible without meeting on your dive boat."

The captain shrugged. "I don't think we deserve that much credit. But we are both so happy for you."

Liv held her hands grandly. "We wanted to treat you to something special. We love the food here but couldn't find the perfect view. And then we had the most incredible idea."

Kasemchai held his hand open toward the door. “Hop inside. It’s a very special restaurant, so we must get there before it’s too busy.”

The driver helped everyone into the car. Liv and Haruka immediately started chatting about Haruka’s wedding and her children. The captain and Kasemchai grinned at each other.

“I love your car,” the captain said. He ran his hand over the armrest under the window, and a reading light beamed in his face. “Sorry, I’m sorry,” he blurted out.

Kasemchai chuckled. “Don’t worry, captain.” He pointed to a small glass panel on the captain’s right arm side armrest. “Tap that panel two times with your finger.”

The captain did and a blissful grin came to his face. He made an *ahh-ahh-ahh* sound as the massage chair feature worked up and down his back.

The car drove along the western coast to Sunset Beach on the island's northern tip, chasing the sun as it started to make its way toward the horizon. At last, they came to the end of the road, and the group exited the car.

The captain rolled out and stretched his arms. As he yawned, he turned to Kasemchai and said, “I don’t understand. There’s no restaurant here.”

“Yes, I know, captain.” Kasemchai slapped him on the back. “But that doesn’t mean we can’t have an amazing meal.”

One of Kasemchai’s attendants stood near the thick green jungle that was hiding the sea from them. He beckoned, turned, and disappeared into the lush tropical foliage. The leaves and branches were cut back to create a shaded path. At the tunnel's far end, it opened up to reveal the sandy beach and clear blue ocean.

The right side of the beach stretched out endlessly and disappeared into the island's far end. The left side of the beach ended abruptly at the base of rugged fossilized coral foundations that leaned out over the ocean—adorned with sago palms and hibiscus bushes that flaunted their saucy rouge flowers. A path of lanterns casting a warm light led to a sunken lounge in the sand. It was lined with soft cushions arranged around a low table in the center. Four chefs stood stiffly off to the side behind a table filled with assorted bowls and boxes of fresh ingredients.

The captain let out a long whistle and finally said, "Wonderful." It was an understatement.

Haruka turned and hugged Liv and then Kasemchai. "It's beautiful."

Kasemchai turned to face the sea. He shoved his hands into the loose pockets of his pants and drew in the salty tang of the air through his nostrils. "The most beautiful place on this island is the ocean, Haruka."

"Good conversation makes it even better," added Liv. "Come on, let's have a seat."

They settled into the soft cushions. A server came towards them, kicking a little sand with each step. She held a black lacquered tray with four silver cans. The word *Orion* was in bold print.

"Ahh, the finest beer," the captain said.

Kasemchai raised his can, and so did the others. "Only the best beer for my friends!"

The captain shouted, "Kampai!"

But it was translated to "Cheers" for Liv and Kasemchai. They crashed their cans together, took a deep swig, and, in perfect synchronicity, they shared a long sigh and looked out over the ocean. Smiles all around.

One of the waitresses brought out a dish of umibudo served on a large shell with a tiny ceramic bowl of citrusy ponzu. In English, they were known as sea grapes, but they were really the size of a pea.

Kasemchai picked up a strand, dipped it into the sauce, and crunched the translucent green bubbles between his teeth. They popped in his mouth, and the slightly briny juice tasted of the ocean.

Behind them, the chefs sliced, tossed, and sprinkled. The waitresses brought plate after plate and filled the table with exquisite-looking sliced sashimi and delicate flavor marriages like mushrooms and shiso leaves with their surprisingly minty astringent taste. Some of the plates sizzled with delicately marbled beef, pork belly, and duck gizzards. Kasemchai fell over laughing when the captain fawned over a plate of sliced butazeri—an acquired taste that he had never acquired.

Kasemchai and Liv shared stories about Sol-vang. With hindsight, their struggles made everyone laugh. Haruka and the captain talked about dive trips and other guests, the changes to the island, and how they had heard about the electric ships and wondered if their dive boat would be electric one day.

Haruka leaned over and placed her head on Liv's shoulder.

Kasemchai smiled and raised his glass. "We changed the world!"

"Thanks to the giant manta rays that brought us here!" Haruka exclaimed.

The captain thought for a moment and then added, "You can't change the world without changing the world."

Everyone fell over laughing. It sounded so absurd and so true at the same time.

The last tangerine bands of the setting sun slipped into the cobalt blue sky. Candles stood tall above the empty dishes, like a lighthouse after a raging storm. A warm glow danced across each person's face.

The conversation came to a lull, and they settled back into their seats to massage their swollen bellies. Remnants of the meal were cleared away, and in their place appeared a set of colorful Ryukyu glassware. The walls of the glass were textured, with foamy bubbles suspended inside with ribbons of wispy green merging with aquamarine. The slight imperfections added character and masked their craftsmanship.

The captain stared at the jungle and then back to Kasemchai. "Time for awamori."

Kasemchai barely moved. "Is it time already?"

"More than you know," he replied.

A movement came from the jungle. Kasemchai looked up and saw that Mr. Tanaka ambled toward them. A young man walked beside him, helping him to walk on the shifting sand. It was the master distiller he had spoken to when he was last in Ishigaki Jima.

He looked at the captain and Haruka.

The captain laughed. "He told me he wanted to meet you again, so I let him know we were here."

Kasemchai stood up and motioned for the distiller to join them in the sand.

Mr. Tanaka held up his hand in protest. "No, no. It's very late for me. I just wanted to share a special bottle with you."

"I can't take any more from you. You've already shared so much with me, Mr. Tanaka."

He touched Kasemchai's arm and gestured for him to sit down. "I won't be staying with you for long." He looked down at the captain, then over to Liv, and smiled serenely. "You have been a very loyal customer, Mr. Seng."

He leaned over and placed a ceramic bottle down on the table. "Don't be misled by this common bottle. What's inside it is very special. When you and I first met . . ." He gingerly knelt on the ground beside the table. It was not an easy move for him, and everyone leaned, prepared to jump up and help him if he fell over. Finally, he straightened. "When you and I first met, you tried to buy my distillery."

Kasemchai looked down sheepishly.

"What's inside this bottle is not so very special. It's what it represents." His frail hands shook slightly as he peeled off the square cloth bound over the top. "This is a batch of awamori made on that day. I was very angry that someone would try to take away my life's work. But something happened that day."

He looked at Liv, leaned forward, and poured her a glass of awamori. He gestured for her to add a few ice cubes with his hands, which she did. "You went diving with the captain and Haruka. I understand there was much dancing and laughing that night." He chuckled, then poured a glass of awamori for the captain and Haruka. "You didn't buy my distillery on that day. Instead, you bought awamori from me every year and shared it with your friends."

He looked over at the young man next to him. "And you gave my grandson a future he can be proud of."

The young man next to him beamed. He turned and bowed to Kasemchai, who raised his hands and clasped them in front of his face. Kasemchai bowed toward the boy and then toward Mr. Tanaka.

Mr. Tanaka leaned forward and poured a glass for Kasemchai, then gestured for him to add ice. "Mr. Seng, this drink is special—because you didn't take the expected path. You found a new one, and we're all better for the way you led us."

He placed the bottle on the table, his hand firmly grasped around the neck. With a surprising enthusiasm, he shouted, "Now, please, drink up!"

Everyone raised their glasses, shouted cheers in their mother languages, and drank.

Mr. Tanaka began to get up, and the young man rushed forward to help him. When he was finally standing, he steadied his feet into the sand and bowed. "Please enjoy yourselves. I hope you don't mind if this old man returns to his home now."

Everyone stood up. They expressed profuse gratitude and apologized for keeping him up so late. Mr. Tanaka grinned and shooed away their compliments with his hand. Then he turned, and with the young man's help, he made his way back to the jungle.

As they shuffled across the sand, the young man turned around and said, "Thank you, Mr. Seng. For everything."

Kasemchai raised his glass to the boy, then the others, and they toasted again. It was very peaceful as they stared at their glasses. The awamori was smooth and had a somewhat sweet taste with a hint of floral and herbal notes.

The captain and Haruka sat down. Kasemchai walked towards the water and stood at its edge, letting the gentle ocean roll up the beach and onto his feet. It tickled his toes and pulled at him as it retreated back towards the open sea.

Liv walked up. She leaned into him and nudged his back with her shoulder. "What's on your mind?"

He smiled down at her. "I was thinking about that boy. How his grandfather wanted to protect his business so that his grandson could follow in his footsteps." He gazed out to the ocean. "Can you believe that bastard refused to sell to me?"

"I can't believe it. There must have been something very interesting for the amazing Kasemchai to be distracted from his business."

Kasemchai looked back at Liv. "There was." He let out a long sigh. "Why doesn't my father want me to be a part of the business?"

"What do you mean?"

"Look at how Philippe has brought Angélique in." Kasemchai stared at the warm ocean rushing over his feet again. "He's never done anything like that for me. I've tried to show him how I should be a part of his negotiations with the Saudis, and it's gotten me nowhere."

She rubbed her hand in slow circles on his back. "You're missing the point. He probably didn't want to force you to be a part of his business so that you could pursue your own dreams. Why don't you just ask him?"

"Just ask him?"

"Yeah. Don't tell him how you're going to solve his business. Just ask him if you can help."

Kasemchai put his arm around Liv and gave her a squeeze. It was strange. He felt very close to her, but they had developed a professional relationship. Would he ruin everything if he tried to kiss her?

He looked back at the captain and Haruka. He knew he would sleep alone again tonight. What a strange island. "You're right. I'll do it."

Chapter Fifty-One

SKIN EXPOSED for two seconds to water heated to one-hundred-and-fifty-two degrees Fahrenheit would result in third-degree burns; a burn that destroys the first three layers of skin and fatty tissue. Paradoxically, because air is less conductive than water, if you were in a room that was one-hundred-and-fifty-two degrees Fahrenheit you could survive for an hour with no burns.

Outside, under the relentless sun, things get complicated. The ground would be hotter than the air, melting the soles of your shoes. Water would evaporate instantly, making your mouth feel parched. You'd feel the need to drink almost immediately. Your breath would become labored, literally drying out your insides with each inhalation of heated air.

At 11 a.m., the desert sun beat down like a relentless hammer. Four men stood in the shade of an oil rig platform, preparing for their shift change. The rig was on the northern edge of the Ghawar oil field—the most prolific oil field in Saudi Arabia, on the coast of the Arabian Gulf. The temperature was one-hundred-and-fifty-two degrees Fahrenheit.

Jacob was the driller. He oversaw the drilling operation on the rig platform and monitored settings like drill speed and the Kelly System gearbox. Foreign nationals performed most of the labor in Saudi Arabia.

He was the only man on this crew from the Philippines. All the others were Egyptian.

Jacob tapped the assistant driller, Omar, on the top of his helmet. Everyone had a mic and speaker in their helmet, but still, he had to shout over the never-ending drone of the drilling rig. "You ready, kid?"

"Yeah. Thanks for letting me stream this. I want my friends to know there's still work to do out here." He commanded his AugWatch through his helmet mic to begin recording. On the top of his helmet was a camera in a solar shield to protect it from the sun. A red light turned on next to the lens. "Hey, folks, I'm here on the rig deck today. I want to show you what it's like to work in extreme heat. You know it's hot, but it's also fun working with a great crew."

He turned and faced the two men next to him. "This is Karim and Darius. They are the roughnecks."

The two men raised their hands to wave as best as they could in their thick suits. With the gold visors down over their faces, they looked like astronauts on Mars. Originally white, Karim and Darius' suits had turned mostly dark grey from the drilling mud that had splashed up from the borehole.

"Tell me about the suits," prompted Omar. "How do they keep you cool?"

Karim laughed. "I don't know if 'cool' is the right word in this shit weather." He looked up at the sun and then down again. With his gloved hand, he patted his arm. "There are tiny tubes in the fabric filled with chilled water that's constantly circulating. I can still feel a little radiant heat between the tubes, but they work pretty well."

"Yeah," Darius said. He stomped his heavy boots on the ground. "But we work hard, and we still sweat. I could swear all the sweat runs down my body and pools in the boots. It's swampy down there."

Omar peered down at the roughneck's feet. "What about the boots? They look like they're rubber."

"Yeah, they're rubber, a special kind that can withstand more than five hundred degrees Fahrenheit. You know, like oven mitts."

"That's amazing," Omar said. He looked up at his monitor inside his shield. "Hey, guys, six hundred and fifty-two people are watching you right now!"

"No kidding," Darius said. "Check this out, folks!" He spun Karim around to show off the pack on his back. "Oxygen right here. The air's so hot you can't breathe it. But the suit is completely self-contained. Totally safe inside here."

Omar turned to Jacob. "And what does the driller do?"

Jacob opened his mouth when a pulsing beep sounded in all of their helmets. "Well, guess I'll have to show you. Let's get on the platform, guys," Jacob said.

The men turned and climbed up the ladder onto the rig platform. Designed for extreme abuse, everything was oversized, metal, and scorching hot. The roughnecks stood over a hole wider than the five-and-a-half-foot diameter of the pipe that emerged from the center.

Omar pointed at the stack of pipes in the rack. "Each pipe is thirty feet long. Right now, they travel seven thousand feet underground and weigh over a hundred thousand pounds. At the bottom is a spinning drill bit, designed to rip through earth and rock."

He pointed to the hole in the platform where the pipe slowly rotated. Suspended from cables above was the Kelly drive—a massive block of steel with hoses and swivels that drove the rotation and downward force.

Darius swung a heavy tong into place, latching it onto the pipe. The tongs, oversized wrenches suspended by cables, gripped the pipes to screw and unscrew the sections. Heavy chains fed off to the side to keep them from spinning wildly out of control.

The Kelly drive stopped pushing down and lifted the pipe for the next section. Just as the join emerged, Karim dropped the rotary slip into the gap, its teeth biting into the pipe and keeping it in place. Darius and Karim worked in practiced sync, swinging the tongs in opposite directions to break apart the joint.

Karim watched the long cables above the Kelly drive that disappeared at the top of the rig tower, and Darius was staring at the joint, getting ready to grab it when the pipes separated. His eyes shot open as he watched the joint lurch up two feet. He never saw the rotary slip pop out of the hole at his feet.

Seven thousand feet of pipe spun and began to gyrate out of control. The heavy tong that Darius held snapped free of the safety chain and whipped around. In a fraction of a second, it reached Karim and smashed him in the chest, throwing his body across the floor. He was still alive, but only for a few minutes as his crushed chest tried to breathe the hot air that invaded his cracked helmet.

The mic on Darius' helmet picked up a long, hollow, twanging sound. Puzzled, he looked up to see the cables slapping overhead. He stared, trying to make sense of it in the seconds before the entire Kelly drive and thousands of pounds of cable slammed down to the earth, crushing him.

The camera on Omar's helmet was enveloped in smoke, then pointed to the ground as he dove for cover. One thousand three hundred twenty-three people were watching the live feed at that moment.

One thousand three hundred twenty-three people immediately shared that video.

Chapter Fifty-Two

MAKOA TAPPED on his AugWatch when he saw the headline, "Ports Brace for Violence."

> Protesters had become more aggressive in major ports worldwide. The group of men and women known as "tankermen" ran the bunker barges that supplied ships in port with the fuel they needed for their next voyage. They were protesting the rapid expansion of electric ships. The tanker ships ran on batteries instead of bunker fuel.
>
> In the Port of New Jersey, a group stopped a truck carrying Solvang batteries and aggressively pounded on it with their picket signs. Police arrived and safely escorted the truck into the port, but not before the group had spray-painted on the side of the truck the phrase, "Batteries kill jobs."
>
> "They need to wake up and look around them," said Steven Tyler, a trucker who was held up by the protestors for over an hour. He said that trucks long ago switched to battery power, and it was only a matter of time before everything was electric. Asked if he would support the protesters if they didn't block his entry to the port, he said, "We call it the Tankermen Tantrum. They need to accept that the world is changing."

After months of protests, some ports had seen groups from other areas joining the protesters. In the Santos Port in Brazil, the oil workers threatened to withhold fuel entirely until the electric ships stopped docking. The Seafarers International Union was working diligently with both sides to prevent a slowdown of the US ports.

Republican senator Tiffany Day expressed frustration that the workers were standing in the way of progress. "No one has said their jobs are in jeopardy. Electric ships produce significantly less waste. These protesters are only afraid of innovation. They're Luddites."

Are tankermen Luddites? *Click this poll to see what readers like you think.*

This is an AI-generated article based on the following sources: *A video posted by JalenLegend8. A video posted by*. . .

Makoa angrily swiped the article away. He jumped up from the couch and paced around the living room before whirling to face his wife. His voice trembled, but his words were venomous. "What did I tell you! It's only a matter of time until I'm laid off again. We're going into Christmas and still can barely pay our bills. I'm tired of always being behind. Always being the loser."

"It's fine, honey. We'll be fine." She leaned forward on the couch and twisted her hands round and round. "The girls are both doing well in school. We have our health."

"It's not fine. You heard what she said. 'No one's job is in jeopardy.' That's exactly what they are: in jeopardy." He pounded his fist on an imaginary table. "And the 'Tankerman Tantrum'—what the hell is that? All they see are protests, and no one is asking why!"

Chapter Fifty-Three

KOSAL LED Kasemchai down the hall to the library.

His father was seated on a long couch that ran under the windows, turned at the corner, and ran along the other wall. Filtered sunlight passed through the thick wooden shutters. Kasemchai walked across the auburn-colored teak floorboards towards his father.

Preap held a VidSheet in his hand. He was so engrossed that he didn't look up for a few awkward moments. When he realized he was being watched, he nearly dropped the VidSheet.

"Why do you call this a library?" asked Kasemchai.

Preap grinned sheepishly as he placed the VidSheet down on the table beside him. "Because I read in here."

"But there aren't any books in here. Just the VidSheet. You could read anywhere."

"Well . . ." He looked about the room with an unhurried gaze. The tension around his eyes relaxed. "A library is a place to read. It's not what's in it, but why I'm here."

Kasemchai sat down on the couch beside his father. "You should call it a reading room instead."

"Perhaps. But you see, it's this room that unlocks my library. The books are all in my head. When I enter this room, the books come back to me, and I read them. That's why it's a library and not simply a reading room."

It sounded like a silly answer to Kasemchai. His father just liked the sound of a library and probably believed that was the kind of room that would be in a palace. As usual, he had come up with a reason to support his belief—one where the shelves of books were imaginary.

"How's business?" asked Preap.

"It's okay. We've converted three hundred more ships in China." He thought a moment. On the way over here, he had thought of a thousand different ways to bring up the subject and, in the moment, decided the direct approach was best. "Father, I want to be a part of the family business. How can I help?" He mulled the question over. And then he added, "We have a big challenge right now."

"What's that?"

He realized immediately he had buried his request to join the family business by adding his own problem, but there was nowhere to go now but forward. "It's called the Emerald Initiative. The Saudis have said they're raising funds to produce batteries for ships. Direct competition to our business."

"Why are you worried about what they might do?"

"Because the Labonnes have already slowed down on new orders."

One of Preap's eyebrows arched. "Why is that?"

"Father, you know the reason. They're getting oil for such a low price it doesn't make sense to convert ships anymore."

His father smirked. "And why do you think I would know that, Kasemchai?"

Kasemchai could feel the anger rise within him. He wanted to tell his father what he knew to be the truth—he and Philippe had plotted all along to use Kasemchai's business to lower the cost of oil. Something deep inside him begged him to hold back.

Kasemchai exhaled and said, "I spoke to Angélique. She told me that they were holding back on new orders because the threat of our batteries had driven the price of oil down. Some sort of deal they made related to the Emerald Initiative."

His father just stared at him, which was maddening. He realized he was about to be made into a fool. He was always too direct when debating. Secrets were leverage in negotiations, and he had none left.

"I did not make any such deal with the Labonne family if that's what you're driving at." Preap crossed his leg over the other with a little assist from his hand. "It's true what you say. The Saudis are subsidizing the cost of fuel for us. I didn't agree to convert my ships as a negotiating tactic. But I did tell you that your battery business was dangerous. We need oil to ship water." He paused for emphasis. "There are consequences when we start switching our ships to battery power, and this is one of them."

Kasemchai pushed his fingers into his kneecaps as hard as he could. He glowered at the floor. "You didn't make any deal with the Labonnes on the price of oil?"

"There's no deal to make. Of course, Philippe and I talk about the price of oil all the time." He leaned in toward Kasemchai. "But we didn't have to push the Saudis to make this deal. They did it on their own."

Kasemchai studied the floor for a moment, then turned to his father. "How did Angélique get so involved with the family business?"

His father leaned back, and laughter burst from his mouth. "Is that what this is all about? That you finally want to be a part of the business?" The grin dropped immediately from his face, replaced by a stern expression. "Kasemchai, you've always been an entrepreneur. Your investments, your innovative ideas—you're so creative."

Kasemchai met his father's gaze with determination. "What I want is for our family to become the greatest water empire. I thought the batteries would demonstrate how we could embrace cutting-edge technology." He lowered his eyes and shook his head slowly. "But instead, you pushed me to sell this technology to the Labonne family first."

His father placed the VidSheet aside and moved closer, placing a reassuring hand on Kasemchai's shoulder. "Who's to say those batteries won't backfire on the Labonnes? I don't understand how the technology works, and I don't know what this change will do to all the systems we have in place. The shipping industry has been oil-dependent for well over a century. Consider how many people it would displace." He grinned and looked out the window for a moment before facing his son again. "I recognize your exceptional creativity, Kasemchai. And I have a proposition for you."

Kasemchai observed the soft outer edge of his eyes and thought how his eyes were starting to soften like his father's. Was it kindness or weariness?

"What's your proposition?"

"How about you work with me as our Chief Technology . . ." He cocked his head up to stare at the ceiling. A smile crossed his face, and he looked back at Kasemchai. "Our Chief Technology Visionist. It means you will need to really figure out our systems. Spend some time understanding how things work before you start to change them."

Kasemchai was torn. As a child, they often spoke about him taking over the business. With time, those conversations faded. Now, once again, he was reaching out to Kasemchai.

But a Chief Technology Visionist? What the hell kind of made-up name was that? Would he report to the CTO or his father? It suddenly occurred to him that his father was still waiting for an answer.

Kasemchai put on a proud face. "Sure. Chief Technology Visionist. When do we start?"

Preap's face brightened immediately. He leaned forward. "Let's start next week. I'll schedule time for you and me to talk about a few things. We'll start scheduling meetings with key people in the company. That should give you time to separate from Sol-vang."

Kasemchai's head whipped up. "Sorry, separate from Sol-vang?"

"Look, I get it. You're still an investor, and you need to monitor your investments." He wagged his hand with his index finger extended. "But if you really want to expand our empire, you need to free up time to focus on our business."

His father stood up and put out his hand. Kasemchai stood, and they shook hands firmly.

Kasemchai exited the room as quickly as he could. The pride on his father's face made him queasy. What kind of stupid name was a Chief Technology Visionist? It was a name that said, "I don't know where to put you, but you're family, and I have to give you a job."

He stormed down the hall, consumed with thoughts of what to do with Sol-vang. What to tell Liv. What would Angélique think of his new title? Would she laugh in his face directly or later, behind his back?

Slowly, his mind returned to the present when he realized Kosal was approaching him. He was carrying a small tray with a cool drink for his father. Kosal had worked with their family since they moved into their first home after the flood. He had slept on the hardwood floor outside Kasemchai's door with nothing but a small pillow.

"Kosal, I have a question for you," Kasemchai said.

"Of course, sir."

"If we all died, what would you do?"

The fear on Kosal's face almost made Kasemchai laugh. "That would be dreadful, sir."

"No, no. It's a hypothetical question, Kosal. If you had to start over, what would you do?"

"I guess I would find another family to work for. I would be their servant."

"That's not what I mean. What if the world didn't need servants anymore? How would you start over?"

"Well . . ." Kosal pondered the question. "I guess I could be a waiter."

"Really?" Kasemchai knitted his brow and scratched his head. "You're so talented, Kosal. It's like you read our minds and you always know what we need. How could you just become a waiter?"

"I'm sorry, sir. The truth is, I don't have the skills for anything else."

Kasemchai nodded. He was thinking about the answer. "That's what I wanted to know, Kosal. And don't worry, no one is going anywhere. Your job is safe."

Kosal blew out a sigh of relief. It didn't come from his mouth so much as deep in his lungs. "Thank you so much, sir. Will you need anything else, sir?"

"No, I'm good. Thank you."

Kosal bowed and scurried down the hall towards his father's office. Kasemchai turned his head to an awkward angle and watched him disappear. Then he straightened and brought his AugWatch up toward his face.

"Chanthavy, call Liv."

After a moment, Liv's avatar appeared. Kasemchai turned and began walking down the hall again. Liv's avatar walked with him.

"What's up?"

They passed a photograph and several large paintings as they chatted. "Who are we putting out of work?"

"Excuse me?"

"I was thinking about what the captain said. You can't change the world without changing the world."

He stopped when Liv passed a smaller painting of a fishing village on the Tonlé Sap. The artist's brushstrokes were thick and textured, with bright, saturated colors. The sun was rising in the background and on a canoe in the foreground, almost in silhouette, was an Apsara dancer wearing a golden headdress and pleated silk dress.

"I think he was referring to unintended consequences," Liv said.

"Right. And when you retrofit enough ships, someone's job will be in jeopardy."

"That's a good point. Well, anyone who works in the engine room, anyone who works in the fuel industry, and anyone who fuels ships, the tankermen."

"And what jobs are we creating?"

"Lots of jobs. There's the people retrofitting the ships, swapping the batteries, running the ECs."

"Great. How do we get the people from one job into the other?"

"They're not interchangeable, Kasemchai. Completely different skill sets."

Kasemchai's eyes grew wide with excitement. "Exactly! We need to train them."

Liv looked like she was about to say something but then changed her mind. Finally, she spoke. "We are the innovators. You've said this again and again, Kasemchai. We need to think big. It's not our job to help people change. That's what the government is for."

"When I was a kid, Liv, my mother always said, 'Leave the room cleaner than you found it.' She was probably talking about the bathroom—or maybe the kitchen. I was a messy kid—"

He studied the dancer. His father always stopped and, with a wispy look in his eye, told Kasemchai that his mother used to be a dancer.

"Kasemchai," Liv interrupted him.

"Sorry, the point is. We're changing the world in a big way. I know you haven't been in your position for a terribly long time. But I'm certain you know that technology is the easy part. It's people that's hard to get right."

"Okay, okay. So what are you going to do?"

"Me?" Kasemchai laughed. "I'd like you to pick a team to run this. I'll let them know what I'm thinking."

Chapter Fifty-Four

ALTHOUGH DALIAN WAS BEST KNOWN for its shipbuilding, the majority of the area inland had been farmland before the rain stopped. After the topsoil blew away, the wind slowly polished the land, making it hard and barren—the perfect place to store large ship engines.

Xiaomeng had been the storage facility manager for two years. It was a lonely position watching acres of land covered in engine parts. At first, they filled only a small corner. Now, hundreds of engine blocks from the Seng family were stacked in neat rows. They towered overhead, the tallest forty feet tall. Stripped of all their wires, hoses, and pipes, they were exposed and skeletal.

Rows of pistons, nearly twice the height of Xiaomeng, reached up to the sky like trees. The piston crowns, with their many compression rings, cast a welcome shadow. They were smooth and slightly blued from the constant explosions in their former role.

Xiaomeng wore his bright yellow-green vest with reflective stripes—a safety precaution required by the company, not that there was any danger that anyone would run into Xiaomeng as he wandered the desolate graveyard.

Today was a nice change, though. Every three months, the company inspector visited and walked the fields with Xiaomeng. It wasn't just the welcome company, although he did appreciate the opportunity to speak with another human being. The men had bonded over a shared burden. Both men had daughters with leukemia. Xiaomeng didn't enjoy talking about his daughter's leukemia; it raised so many painful thoughts, but he couldn't stop talking about it. With all his heart, he believed he would help her overcome this cancer.

The two men strolled side by side. They wore broad hats to protect them from the sun. YanYan held his wrist with his AugWatch up and watched the visuals jump from part to part. Arrows and highlights called out areas on the engine parts that should be inspected visually.

YanYan pointed to one of the white plastic covers that had fallen off a port and was lying on the ground. "Make sure the port covers are placed back onto the engine. It's okay if they're loose, but we don't want them blowing away."

Xiaomeng picked up the sheet of plastic and identified which port it had come loose from. He didn't have any tape, so he stuffed it into the hole, stood back, and marked the location with his AugWatch. "I'll get some tape and return here later."

"Good. Good." YanYan took a cloth from his pocket and wiped his brow. He turned back toward the building. He couldn't see it from this distance, but he knew it was there. "These walks take longer and longer every time. How many engines since I was last here?"

Xiaomeng asked his companion. He gave a soft laugh. "Two hundred and thirty-seven. More miles to walk. It's how I stay in such good shape, you know."

The two men started meandering their way back to the building.

"How's Jia doing?" asked YanYan.

Xiaomeng sighed. He looked up into the sky and then down again. "She needs a bone marrow transplant."

"Huh, I see."

The two men walked in silence for a moment. It was warm, but the two men were always in the shade of the towers of machinery.

"Is it scheduled?" asked YanYan.

"No. We have permission to get the transplant. As you know, it's risky, but it may be her only hope."

"But that's not the problem, is it?"

Xiaomeng smiled wanly. "I would, of course, give her everything I own to help with recovery. But the long-term care costs so much. What if I have no more to give? She could die in recovery before the leukemia would take her life."

"Hmmm," YanYan said. "My Yuting is doing well. She has many days that are . . ." He gazed off to the side. His eyes watered, but he held off the tear that wanted to break free. "There are hard days. But she is recovering."

Xiaomeng gave the other man a comforting squeeze on his arm. "I'm happy for you."

They turned down an alley between rows of auxiliary diesel generators stacked two high.

"Xiaomeng, I could not afford the treatment for Yuting."

Xiaomeng stopped. The center of his brows scrunched up as he regarded YanYan with concern. "What do you mean? Do you need money?"

"No, my friend, I am fine. But I know that you need help. I've been thinking about this for a long time. I know how we can pay for your daughter's medicine and all of her recovery."

Xiaomeng's face contorted from surprise to fear to confusion. "What do you mean?"

"I want you to do something. Each week, I will have a canister delivered to your home. I want you to start at the back of the field, all the way at the end, and place the canister on the ground and walk away."

"What is in this canister?"

"It's a gas called hydrogen sulfide." He held his hand up in front of his face. "Don't breathe it because if you breathe it for a long time, it is poisonous. But you won't be anywhere near it. You leave it on the ground, press a button, and the gas will begin to escape from the canister."

"This sounds like a dangerous story you are telling me, my friend." He shot him a stern glare, silently imploring him to stop.

YanYan ignored him. "Every week, you will let this gas escape, and the next week, when the gas is long gone, you pick up the old canister and place a new one in another location."

"And what does this gas do?"

"When it mixes with the moisture in the air, it creates an acid that can damage the protective coating on the parts and start to eat away at the metal."

Xiaomeng stopped and turned to his friend, eyes wide. "You want to destroy these engines? Why?"

"They won't be totally destroyed, but they will be damaged. It will be very hard to put the engines back into the ships. And they may never pollute the air again."

"You are an eco-terrorist?"

YanYan looked off into the distance. "No, I'm not an eco-terrorist. I'm a father like you. Every quarter, when I return, I will inspect your progress. And every quarter, you will get a payment."

The two men turned and faced each other.

YanYan placed his hand on Xiaomeng's shoulder. "Get Jia's bone transplant now. There's no time to waste."

Xiaomeng's eyes crinkled as a small tear rolled down the corner of his eye. The mention of his daughter's name vanished any doubts or concerns he had about destroying the Seng family's property. He knew that he had only one path forward.

Chapter Fifty-Five

Makoa pushed open the door and stepped into the office. The office manager, Fran, was at the front desk, awkwardly tucked in an alcove behind the front door.

She yelled out to Makoa in an unusually loud voice, "Morning, Makoa. Are you thirsty?"

"Hey, Fran," he said without slowing down. He walked straight to Peter's office in the back. It was sparse, a metal table with an overstuffed chair that squeaked every time the owner of the company, Peter, leaned back. An old picture of wild horses in a meadow, possibly left by the previous owner, hung between the faded rectangles where pictures hung in the past.

Ben and Peter were in the office. Ben was leaning over the desk with his palms placed in front of him.

"It doesn't matter," Peter said. "I know you don't want to, but that doesn't change the fact that every day fewer ships need fuel. That means fewer rotations, pure and simple. You have to think about the experience of your crew."

Peter looked up and saw Makoa standing in the doorway. "Oh, hey, Makoa. What can I do you for?"

Makoa glanced from Ben to Peter. He tapped at his AugWatch and pulled up the work schedule. "I wanted to see what was going on. I don't have a shift next week."

Peter started to say something and then stopped. It was clear he was trying to find the right phrase when, finally, he said, "I have to take a barge offline. It doesn't look good for us, and I need to make some tough decisions."

"I don't understand," Makoa said. "I know the fuel price spiked, but it's coming down already."

"That's not it, Makoa. It's not about the price. Every day, fewer ships need fuel."

Makoa's hands were deep in his pockets. He squeezed them tightly, the nails biting into his palms. He couldn't look at Ben. "Will I get a shift the week after?"

"I don't know, Makoa. The thing is, I have to give my senior guys the first shift."

"What about Ben? He's practically ready to retire. Are you going to keep giving shifts to him when you have someone like me to count on for years to come?"

The veins on Ben's neck bulged. "I've still got years ahead of me, Makoa. Besides, you're always talking about electric ships. Why don't you just get a job taking care of them?"

Makoa leaned in. "Why would I do that? I've been trained to work on a bunker barge. I work here."

Peter slammed his palm on the table. "Cool it, guys. Ben, you need to keep it under control. Makoa, you're still in my rotation. The short of it is I have fewer barges to send out. People get sick, things happen. But you're going to miss some rotations."

Makoa was quiet. The men looked around at each other, waiting for him to say something. He let out a long, slow breath. Finally, he spoke. "I got it. Thanks for thinking of me, Peter. You want me to check in next week just in case?"

"Sure, Makoa. Give Fran a call. She's always there to keep you in the loop."

Makoa spun around and walked out. His shoulder slammed into the doorjamb. It wasn't intentional, and it hurt. His face flushed as he realized it made him look like he couldn't control his temper or that he was clumsy. He stormed past Fran and gave her a quick nod goodbye.

Outside, he stopped and stared up at the stars. He closed his eyes and muttered, "Shit." He looked down, kicked the dust on the ground with his heavy boot, and started towards the bus stop. "Shit! Shit! Shit!"

IT HAD BEEN three long months since he last had a shift, and the days seemed to blur together. Makoa sat on the edge of his bed, staring blankly at the wall. The room was dim and stale. With the door closed, the room didn't cool off, and he ran his hand across his temple when the sweat had swelled enough to run down the side of his face. Sometimes, it ran into his eye and stung, but he didn't care.

He didn't even watch *Look Up to God*. Sometimes, his heart raced when he thought of all the things he needed to buy for his family, but mostly, his mind wandered aimlessly.

His two young daughters, full of energy, ran through the house. They had learned to be quiet when they passed his door. They didn't know what was wrong with him, but they didn't like it when he yelled. Or worse, when he said nothing.

When he wandered through the house, they tugged at his sleeve and called out his name, but he barely noticed them. He shrugged and walked away. He had something to do but couldn't remember what it was. In the cramped space, there was nowhere to escape, and the confinement only amplified the sense of despair.

On Sunday, a post caught Anela's eye. The girls ran through the living room and begged to go onto the porch. She gave them a bit to eat and opened the door for them. She looked again at the post on her AugWatch, went to the bedroom, jumped onto the bed, and pulled the blanket back from Makoa's face. He yanked it back just as quickly.

"What are you doing?" came a muffled grumble from under the blanket.

"You've been moping around for weeks, and you've been in bed for days. It's time to get back up. And today's Easter. I've hidden the eggs, and there's breakfast on the table."

"The girls are too old for hunting for easter eggs."

Anela smiled. "Apikalia is only ten. And both girls enjoy the hunt." She sat up and crossed her legs. She poked at her AugWatch to bring up the article she had found. "Check this out." She pulled back the blanket again and held the AugWatch in front of his face. "Sol-vang has a program that trains people to work on electric ships. And there's a full scholarship for people with experience like yours. You need a TWIC, a Firefighting Course certification, and a Cargo Course certification. You have all of those."

Makoa closed his eyes and pressed his head back into the pillow. "I'm not going to go through more training. Besides, the guys literally said I would go to work on an electric ship. They'll think I'm a traitor."

Anela leaned over and kissed Makoa on the forehead. "They are the ones who aren't giving you any shifts. They're the traitors. Besides, I didn't tell you the best part. The positions are all salary, not hourly positions."

Makoa slowly opened one eye, then the other. He looked so tired.

Anela leaned over and started rolling him toward the side of the bed. "Come on, sleepyhead. Rise and shine."

Makoa slowly swung his legs over. He sat on the bed and stared at his feet on the ground. He looked towards the door. The girls' bedroom was there, just out of sight.

He groaned and rose from the bed. "Okay, let's get this morning started."

Chapter Fifty-Six

GOVERNMENT BUILDINGS HAD ALWAYS HAD a utilitarian aesthetic—an inclination for neutral colors, harsh ceiling lights, and highly waxed tile floors that looked simultaneously polished and lumpy. The only panache in this hallway was a tasteful poster in gold and blue reminding people that Independence Day, September 7, was next week.

Liv and Johannes stood in this hallway waiting to be let into the meeting room in the Brazilian Ministry of Defense Headquarters. A soldier in head-to-toe khaki opened the door and ushered them in.

Liv glanced around, trying to read the mood in the room. The Secretary of Defense Products was all smiles, while the Secretary of Defense Policy had heavy eyelids that drooped at the corners. He gave no indication that he recognized they had entered the room. The Minister of Defense looked up, frowned, and looked down again. The others in the room appeared as if they were trying their hardest not to show any expression whatsoever.

Johannes had spent months negotiating with several members of this group. In fact, it was Lucas, the Secretary of Defense Products, who suggested they create a joint partnership with Sol-vang and China. But it was clear that Liv and Johannes would need to put on a show for the Minister of Defense. She needed to find just the right message that would convince him that switching to batteries would not only revolutionize Brazil's military but also enhance its role as a supplier to the world.

Lucas introduced Liv and Johannes. He reminded everyone that producing armored vehicles was essential to Brazil's security, and exporting armored vehicles was essential to their GDP. He spoke about Brazil being the largest exporter in Latin America and the eighteenth-largest exporter in the world. He started to list all the countries they exported to when the Minister of Defense cut him off.

"Please, we are all aware of the countries we export to, Lucas. I'd like to hear from our visitors."

"Yes, sir," Lucas said. "Let me bring it back to the most salient point. In any operation around the globe, the combat supply lines that fuel fighting vehicles are especially vulnerable to attack. Reducing the reliance on moving large volumes of fuel will"—he raised his hand and pointed with his index finger for emphasis—"save soldiers' lives."

He nodded towards Liv and Johannes. "Sol-vang has transformed military electronic vehicles. Liv Anselm, the president of Sol-vang, has led the development of non-volatile batteries with their revolutionary swappable battery system. Johannes Nielsen, head of government and military applications, has worked closely with my team to explore a valuable partnership."

He crisply raised his hand to Liv and Johannes and stepped backward so they could come to the front of the room. Liv spoke first.

"Ladies and gentlemen, we have a proposition that might sound audacious, given the history of oil-dependent military operations. However, it is through bold actions that we can achieve profound transformation. We understand your skepticism. Oil has been a reliable source of energy, but it's not without drawbacks—volatile prices, environmental consequences, combustion risks, and, most critically, the risk of dependence on a finite resource."

Johannes painted a vivid picture of how Sol-vang technology would revolutionize Brazilian military vehicles, emphasizing the reliability, longevity, and rapid recharge capabilities. Liv and Johannes then presented the details of their partnership proposal. They described how Sol-vang would collaborate with Brazilian firms to establish a state-of-the-art production complex in Recife, a city on the eastern tip of Brazil.

"This complex," Liv said firmly, "will be a testament to the fusion of our technology with Brazilian expertise. Under this agreement, Brazil will not only harness the power of Sol-vang technology but also maintain sixty-six percent ownership of the venture. This is a true partnership that will empower Brazil."

It was at that moment that Liv realized everyone was scrutinizing their AugWatches. She heard Lucas behind her let out a low whistle.

"And the reason . . ." She pressed her lips and glanced at Johannes, whose attention was on his AugWatch. The room was murmuring.

Johannes raised his arm to show Liv the news on his AugWatch. She mouthed the news article's title, "Breaking News: Saudi Arabia in Chaos."

"What does this mean?" she whispered.

"That's not all," Johannes said. "Jihadists have blown up the water pipeline to Dammam."

Dammam was Saudi Arabia's new capital since the drought began.

Liv's eyes went wide. "The country is going to collapse. This is going to impact supply chains in catastrophic ways."

Lucas stepped up and put his hand on Liv's shoulder. He looked across the table at the Minister of Defense and nodded in affirmation. "Liv, Johannes, thank you for your presentation. In light of this news, let's pause here."

Liv and Johannes didn't say a word. They didn't frown, or grimace, or smile.

But one thing was absolutely clear: they couldn't have asked for a better ending to their presentation.

Chapter Fifty-Seven

MAKOA ENTERED the classroom at Barontun University, a mere six miles from the Port of New Jersey. He had anxiously rubbed his left forearm the entire walk to the classroom, and now it felt a little raw. There were a lot of people strolling along the paths, and they were unnaturally happy. They were very young. He didn't fit in here.

In the classroom, he could relax a little. The people in here were all older, like him. They looked callused and worn out, like him.

Three weeks had passed since the course started. He felt sluggish, keenly aware that he was starting over again. Ever since he left Hawaii, his life had been nothing but a repeating loop of starting, failing, and starting again.

The instructor had a small table set at an angle in the corner, which he used mainly as a seat. A rolling table was set up at the front of the classroom with an experience sphere on top of it. He wandered from left to right, stopping suddenly and running up to the experience sphere to sketch out designs as if the lesson was literally being designed at that very moment.

He frequently projected models, turned them, and manipulated them in front of the class. Makoa thought they'd get right into tools, but instead, they spent a lot of time studying how the electrolytes flowed from the anode end of the battery to the cathode. He was starting to get words to describe the batteries that powered everything in his life.

The room had low tiers with rounded tables. Makoa picked a row slightly closer in the classroom today. He leaned forward, ready to begin the lesson. The instructor seemed jumpy and moved more quickly in the front of the classroom.

"Hey, folks, today's lesson will be a little different. When I'm not in the classroom, I often think about why you are in this classroom. Where will you go with this knowledge? How will your life be different?" He walked over to the desk, sat down, then immediately stood up again. "Well, perhaps the best way to understand how this course will help you is to understand why this course was created in the first place. I have a special guest joining us today. Everyone, I'd like you to give a warm welcome to—"

He walked over to the door, pushed it open, and leaned outward. He hung there for a minute, talking to someone out of sight, then stood up again. "Ladies and gentlemen, the lead investor in Sol-vang and the founder of the Sol-vang Reskilling Program, Mr. Seng."

As Kasemchai entered the room, everyone leaned forward in amazement, exhaled simultaneously, and began to clap.

Kasemchai looked a bit sheepish. "Hey, folks. How's it going?"

The instructor held his hands up to quiet the crowd. "Okay, okay. Thanks, everyone. I was a little stunned myself to hear that Mr. Seng would be joining us. Of course, you all know him. Mr. Seng, this is a great honor for us all. Why don't you share with us why you stopped in today?"

"Sure, I'd be happy to." He scanned the room. "Do you have something like a stool for me to sit on?"

"Of course, of course," stammered the instructor. "I don't know why I didn't think of that before you arrived." He never stopped talking as he carried the stool to Kasemchai. "I'm sorry, really, I should have—"

"That's okay," Kasemchai said. He sat, rested one foot on the lower rung, and started. "As you all probably know, the revolution in electric tanker ships started in Rotterdam, in the Netherlands. Now, the idea didn't start as a big idea. We just wanted to reduce the reliance on bunker fuel to power tanker ships. It was really a cost-saving idea that happened to greatly reduce our carbon footprint. What many people don't realize is that tanker ships are one of the biggest emitters of carbon gas out there. And we can't stop using ships. More than eighty percent of all goods travel by ship."

He paused and looked out on the room. Dozens of faces watched him, not because of who he was related to but because of how he was changing their lives. They had a connection to him, and he realized how good that made him feel.

He smiled and continued. "When we started, convincing people to convert just a handful of ships to electric seemed impossible. In a remarkably short time, we turned the impossible into possible. And after the collapse of Saudi Arabia, we passed critical mass. Now, shipping companies are converting their ships as quickly as possible to offset wildly unpredictable oil needed for engines. But a technology company today can't focus solely on its mission to innovate and make money. Technology is inherently scalable, and scalable activities always affect people."

Kasemchai stood and started to walk around in front of the class. "That is why I wanted to stop in today. You are one of the first classes in the Sol-vang Reskilling Program. You are the very people driving innovation, and I want to ensure you have all the tools to succeed."

A woman in the first row raised her hand. "Mr. Seng, do you mind if I ask a question?"

"Of course, ask away."

"Is your plan to make all ships run on batteries?"

"That's an excellent question," he said, crossing his arms as he pondered the question. Then he dropped his arms, with one hand in his pocket, the other in front of him. We want to convert ships to run on a reliable source of energy. The cost of oil is fickle, the source of oil is finite, and the infrastructure is fragile. But there is plenty of room for both fuel and batteries. Companies should switch to battery power when there is a clear advantage."

A man raised his hand, and Kasemchai called on him.

"I just wanted to say, Mr. Seng, that it's very good to hear you say that." He looked around the room, and there were many nods in agreement. "We kinda thought you were going to destroy the competition.

But if you think there's room for competitors, then it just feels . . . healthier."

"Thank you," Kasemchai said. "You're exactly right. Competition is healthy."

Makoa raised his hand.

Kasemchai pointed at him. "What's up, sir?"

"Thank you, Mr. Seng. Let me start by saying how grateful I am to have this opportunity. This training program is very different from anything I've ever done before." Makoa wiggled in his seat. It looked like he was trying to get more comfortable, realized that wouldn't happen, and continued on. "But I have to tell you I still have my doubts."

"What do you mean?"

"Well, the truth is I used to be a winemaker. The winery went out of business. Do you remember me?"

Kasemchai squinted at him. "I'm sorry, have we met before?"

The teacher stood up and looked nervously between Makoa and Kasemchai.

"I worked at the Newark Winery." Makoa's fists were on the table. He clenched and unclenched them. "I was showing you how the robots worked."

Kasemchai grinned broadly. "Yeah. Oh, yeah. I remember you." He glanced off to his side and scratched the back of his head. "It burned down. Too bad. It was a really promising winery."

Makoa shifted in his seat. "Well, the winery burned down, and I was out of work, so I put myself through training to work on a bunker barge. I'm now in training to work on electric ships because my job on the bunker barge fell apart."

Makoa stopped speaking. His story's point didn't seem as clear in his head as when he raised his hand. He felt his chest pounding.

Kasemchai smiled. "Take your time."

He took a deep breath. "Thank you, sir. Well, the truth is, technology doesn't seem to be my friend. The world keeps changing, and workers like me just get pushed from one job to the next. It's constant uncertainty. And the truth is—well, the truth is wealthy people make all the decisions, and people like us . . ."

He paused again and looked around the room. There were no

smiles. The room stared back at him with daggers in their eyes. The guy next to him leaned over and hissed, "Don't piss him off, man."

"What I mean is . . ." He stared down at his hands and took a deep breath. "I have two girls. I want them to have a bright future. I tell them that. But the truth is, the future is completely out of my control. I keep getting knocked down again and again and again."

Makoa wrapped his left hand around the right and rubbed the raw knuckles. "I'm sorry, sir. I really shouldn't have said that. I realize this program is exactly the solution to help people like me."

"No. No, you said exactly the right thing." Kasemchai returned to the stool and sat on it. He leaned far forward, hands on his knees, and said, "Technology is changing our world in dramatic ways—it always has. And the people who command technology generally have the greater good in mind. But technology, by definition, is massively scalable." He sat back and waved his right hand in front of him. "When you look at something from such a high level, it's hard to think about everyone."

He glanced out the door to someone, or perhaps several people, who stood out in the hall. "You know what?" He dropped his hand to his knee and half-snorted, half-laughed. "There are very few people around me who speak as openly as you just have. If you don't mind, I'd like to talk to you more about this. Do you think we could meet tomorrow?"

Makoa looked at the other students in the class. A woman shrugged.

"Yes, of course, sir," Makoa responded.

"Cool. I'll have someone get in touch with you." Kasemchai stood up. "Everyone, thank you all for participating in this program. Like Makoa just said, technology keeps changing, and people get knocked down. But not this time. You folks are at the front this time. I can't wait to see where you go from here."

The room erupted in whooping and clapping. Kasemchai raised his hands and clasped them in front of his face. He bowed and exited the room.

Chapter Fifty-Eight

THE CAR STOPPED in front of a smaller entrance to the tall skyscraper. The overhang shot out like the underside of an undulating river, mirroring the curved blue mosaic that cascaded one thousand feet down the façade.

A man opened the car door and motioned for Makoa to enter the building. He smiled. It was professional more than warm. Makoa walked through the doors and rode the elevator up alone to the penthouse.

When he exited the elevator, someone was waiting for him, and they escorted him toward a long hall. Heavy mirrors hung in panels on his right side. On the left hung a series of paintings with jagged marine blue and green streaks of paint that looked like they had been scraped onto the canvas with a knife. The light at the end of the hall was brilliant and obscured all detail.

The room he entered was expansive. The floor was white oak in French herringbone, and the ceiling was twenty-three feet above, tethered by round white columns.

Kasemchai stood when he saw Makoa. Behind him was Central Park, all two and a half miles, trimmed by a border of tiny buildings.

"Makoa. I'm glad you could make it. Are you thirsty?"

"I'm not thirsty. Thank you." When Makoa put out his fist to bump Kasemchai's, he noticed how much larger it was, but that did not bolster his confidence. "I'm still a little confused about why you wanted to see me."

Kasemchai stood back as Makoa drifted towards the window. It was breathtaking to take in the view of New York from so high. It was 5 a.m., and the sun was rising like a ball of fire. The tall buildings stopped at the park's edge, a rash of nature in the middle of the cement jungle. Most of the trees had been cleared, and new trees had been planted recently, mostly along the paths.

"It's so sad," Makoa said.

Kasemchai jerked his head back, his eyes wide. "What do you mean?"

"The park. This park should be green."

Kasemchai nodded and stared down at the park. "You're right. I've seen pictures from this spot before the rains. It was a beautiful park . . ." He trailed off. "Where I live in Cambodia, it's quite green."

"I imagine it is."

"You know, I wasn't always this wealthy. In Cambodia, my father had a construction business," he said, stopping and looking out across the park. "I think we were really just lucky. We were lucky we didn't die like so many people did that year. And we were lucky we could start capturing the water."

"It's not just luck, Mr. Seng." When he looked at Kasemchai, his face was dead serious. Kasemchai might not appreciate his father, but he did. "Your father did many great things. Sure, he was in the right place at the right time, but he was always in the news, traveling to meet different people and signing agreements. As we suffered, it gave us hope that something was being done."

"I know you're from Hawaii, Makoa. What was the moment when you decided to leave?"

Makoa stared back out at the park. "I was twelve. The water was running out. They told us we had to get on a plane because soon the planes would stop coming to the islands. All the trees had fallen, and the plants just lay down. Everything just stayed where it was, rotting in the sun. The old people wouldn't leave the island, though. They wanted to die there. And they . . ." Makoa caught himself. Those memories of Hawaii were dark. He shrugged. "Why do you want to know, Mr. Seng?"

"My father always said the rain was a gift. Of course, from your perspective, the gift turned out pretty well for my family." His face clouded over. "But it wasn't easy. We also lost many people in the floods. My father had a plan, but I rarely saw my father. He was busy saving the world."

Kasemchai stopped. "Would you like a drink, Makoa?"

"Sure, what do you have?"

"You can have anything. How about a Gibson?"

Makoa rubbed the knuckles on his hand. "What's a Gibson?"

"It's a gin martini. Shaken, over ice. With a pearl onion. The translucent color makes me think of . . ." He scanned the room. "Of purity. It elevates the cocktail."

Makoa glanced at Kasemchai's wrist. "Like your watch."

"What?"

"Isn't that the AugWatch Clear?"

Kasemchai looked at his wrist and laughed. "I didn't think of that, but you're right." He turned his wrist and looked at the watch from the side. "You can tell there's something in it, but it's still totally transparent."

"How the heck do they do that?" asked Makoa.

"No idea, but it's worth the price."

Makoa laughed, then caught himself. The AugWatch Clear was not something he could imagine owning. "Do you have a booch?"

"Of course. Or—" Kasemchai stopped and looked right at Makoa. "Or how about a glass of wine."

His face flushed. Did Kasemchai know about the fire? Was this all a trap to get him to admit to the damage his friends had caused? "No, just a booch."

"Chanthavy, a Gibson for myself. And a hard kombucha for my friend."

Makoa scrunched one eye and tilted his head to the side. "Do you have a robot?"

"A robot? Why?"

"I mean, if you order the drinks on your AugWatch . . ." Makoa trailed off. The conversation constantly put him on his guard and he felt exhausted and anxious.

Kasemchai laughed. "No, it's not nearly that complicated. I just sent a message to my staff."

"Ah. That makes more sense, Mr. Seng."

Kasemchai walked towards the couches, sat down, and gestured for Makoa to sit on the opposite side of the coffee table. He looked up at the ceiling for a moment, then said, "Every year, we celebrate the New Year, right? But for people in Cambodia, we celebrate the New Year in the middle of April. It's called Choul Chnam Thmey."

"Why do you celebrate in April?" asked Makoa.

Kasemchai shrugged. "Why do you celebrate December?"

"It's the end of the year."

"Is it? I suppose that winter solstice occurs in December, but isn't spring when we emerge from winter? A time of birth and renewal?"

Makoa nodded. "That makes sense."

"In Asia, April also used to be the beginning of the rainy season."

Kasemchai looked like he was about to say something, then stopped himself. He returned to his story. "On the first day of the New Year celebration, we clean the house, we make lots of special foods, and we visit the pagodas to give gifts to the monks. On the second day, we show respect to our elders, both to the living and the departed. And on the third day, we wash away bad thoughts to bring good luck.

"I always think on that last day, what are my bad thoughts? One of my bad thoughts was that I deserved this life. That I have somehow earned this wealth."

Makoa wondered where Kasemchai was going with this. The floor at the top of this building was nicer than anything he'd ever seen in his life. But he didn't feel comfortable. He felt trapped.

"The rain is a gift," Kasemchai said. "It could stop as easily as it started. But the electric ships. That is real change. I made them happen. I knew there would be consequences of change so I started the Sol-vang Reskilling Program. Then I met you, Makoa. You didn't simply thank me for the opportunity. Instead, you told me that the world keeps changing, and workers like you get pushed from one job to another.

That evolution brings only suffering for you." His nose crinkled as he grinned. "You basically said that I have screwed up your life."

Makoa was on alert again. "I'm sorry, sir. It's not what I had meant to say."

"But it is what you said. And you said it from your heart."

A servant walked towards them with two drinks on a black lacquered tray. She placed a glass of kombucha in front of Makoa and then the Gibson in a long-stemmed martini glass in front of Kasemchai.

Kasemchai took a big sip. He grinned, then swallowed. "What you said, Makoa, is what needed to be said. I realized at that moment how my technology is changing lives and how you are in transition right now. I need someone like you, Makoa. I need someone who will ground me. Who will tell me the truth."

"Sure. What do you want me to tell you?" asked Makoa.

"No, I don't think you understand." Kasemchai leaned forward on the deep couch. "I want you to work for me, Makoa. Come here once a week. We'll call them dinner chats. I want to understand what the last ten years have been like for Makoa. The good, the bad, and the ugly."

Makoa gripped his knee, deep in thought. "So, you want to pay me to talk to you?"

"Well, yeah, basically. But it's more than that. I want you to be my conscience. Be my—"

"Kasemchai, can you really trust someone you pay to tell you something?"

Kasemchai laughed and raised his glass. "That's exactly why I want to give you a job. You're kind of a jerk, and I need more people like that around me."

Makoa laughed. "Well, if that's what you need, I'm your man."

Chapter Fifty-Nine

LANGUAGE AND GEOGRAPHY WERE TRANSITIONAL, but the Earth's rotation around the sun had been constant for ages. When the sun moved from Pisces to Aries, the first sign of the zodiac, it signaled the start of a new astrological year. It was also the period of the fourth month of the Gregorian calendar.

The fourth month had become known as April, but its original meaning had been lost over time. It might have originated from the Latin term "aperire," which meant "to open." This could refer to the blossoming of flowers and trees during this time of the year.

In April, the bustling city of Krong Battambang came alive as the sun's first rays declared the arrival of Choul Chnam Thmey, the Cambodian New Year. The air was filled with the sweet scent of Champei blossoms and the sound of laughter.

Sarin had spent weeks preparing for this day. She had meticulously cleaned her home, arranged offerings of fruits and flowers at her family shrine, and donned a beautiful silk sarong.

She stepped out onto the street and joined a stream of people who were all dressed in their best clothes. Each of them carried small bowls filled with flower petals and scented water. Together, they walked towards the local temple.

In the temple, the aroma of incense hung thick in the air, and the sound of chanting monks provided a tranquil backdrop. Sarin and others approached the shrine, paying their respects to the elders and pouring scented water over their hands. This gesture symbolized cleansing and purification, a renewal of spirits.

Outside the temple, young children playfully splashed water on one another, their giggles echoing through the streets. Families exchanged blessings, and friends gathered to celebrate with traditional games and dances.

A little boy shouted, "Soursdey Chnam Thmey!" Khmer for Happy New Year.

Sarin turned, recognizing the voice. The little boy ran up and hugged his aunt. She joined his family to visit a famous circus. They marveled at the young men who juggled staffs with fire at either end. They held their breath, watching the acrobats, and sighed, mesmerized by the dancers in colorful costumes.

Afterward, Sarin and her brother's family joined a lively procession that wound through the streets. The procession carried a Buddha statue adorned with flowers and incense. The crowd chanted blessings for a prosperous year ahead.

In the evening, under a sky splashed with stars, Sarin and her brother's family sat down to a grand feast of Cambodian delicacies. The table was filled with aromatic dishes like amok, a fragrant fish curry, and sticky rice cakes. They shared stories, laughter, and their hopes for the future.

On that same day across the border, a bus rounded a corner, and two Thai girls, Nee and Mali, grabbed onto a silver stanchion pole to steady themselves. They laughed out loud as the crush of people surged and threatened to tear them from their anchor. In their hands, they carried bright-colored squirt guns with bulbous canisters for water.

The air conditioner pumped cold air into the cabin, but it was no match for the blazing sun outside and tightly packed bodies inside. Nee tried to squirt her friend to cool her off. As the bus banked sharply, she fired the water gun wildly. accidentally spraying the people around her. Quickly, she raised her hand and called out an apology. The crowd only laughed.

The girls rode as far as they could on the bus toward Khaosan Road in Bangkok. With their water guns in hand, the two friends ran as best as they could through the crowded streets. Songkran was the Thai New Year, and it was celebrated with parades and temple visits. But they were there for the water fights.

They rounded a corner and crashed straight through a riot of colors and laughter. People of all ages held similar water guns, and streams of water burst in every direction. The girls were drenched, their long black hair whipping around and sticking to their faces erratically.

A boy jumped back from a barrage of water aimed at him and knocked Mali over. He didn't notice, but his friend did. His friend bent down and grabbed her hand. He leaned in close.

"Are you all right?" he asked.

Mali looked up at him. She smiled. She landed hard on her bottom, but somehow, the pain had vanished. "I'm . . . fine," she managed to say.

"My friend is such an oaf." He slapped the leg of his friend who was standing nearby. Nee kneeled down, and she and the boy lifted Mali to her feet.

The four stood in a tight circle, panting from running and laughing. Water rained down on top of their heads.

"I'm so sorry," the oblivious boy said. "Are you okay?"

"I'm fine, don't worry."

The boy who helped her put his arm around his friend's shoulder. "I'm Aroon. And this is my friend, Songkran."

Nee laughed. "Songkran? It means 'born during the New Year's festival'?"

Songkran laughed. "Yeah. I was, and my parents couldn't come up with a more creative name."

They stared at each other, gently laughing longer than the little joke. It felt safe and warm inside their circle. No one wanted to disturb the moment. Then, someone jostled Nee, breaking the trance.

"Hey, I have an idea," Aroon said. "Since my friend nearly killed you, we must repay you with a treat."

With water guns in hand, they ran down the streets together. Nee and Mali playfully chased the boys, aiming their water guns in their direction and exchanging laughter and splashes.

They stopped at a booth with bowls of brightly colored fruit and jellies to enjoy nam khaeng sai, a shaved ice dessert. They painted their faces and bought tightly woven bracelets of fragrant jasmine blossoms. They shared music from their AugWatches and talked about their parents and their dreams.

In the evening, the festivities took on a different tone as the Bangkok skyline lit up with fireworks. Nee and Mali, their faces still adorned with paint, enveloped in the scent of jasmine, gazed at the dazzling display overhead. They leaned into each other and laughed. They sighed deeply. No words were needed to exchange the joy they felt at that moment.

On that same day, across the continent, bicycle tires smoothed the dry, muddy bottom of the canal under the heat of the night sky. On King's Day, the canals were a sea of orange, illuminated by thousands of individual lanterns. The fat, tired barcycles were mobile floats, blaring traditional music and adorned with orange banners and flowers.

The streets of Rotterdam were filled with the sound of bicycle bells and laughter as families and friends pedaled their way along. Children with painted faces in shades of orange sat on the back of bicycles, giggling and enjoying the creative and makeshift parade. Vendors offered refreshments and snacks along the route.

In the parks, people gathered for a massive "cycle-in picnic." They parked their bicycles in a circular formation and shared treats and laughter. It was a unique sight as orange-clad individuals maneuvered their barcycles and bicycles while holding plates of fried bitterballen and mini pancakes dusted with powdered sweetener, known as poffertjes.

The last revelers stopped as an orange curtain rose up above the cityscape, chasing away the purple sky. They stared at the brilliant ball that grew as it rose until it hurt their eyes. A wave of heat washed over them, and they retreated to their homes.

On that same day, across the ocean in Brazil where the Rio Negro and Amazon Rivers converge, was a large city named Manaus. Yara was a young girl who lived in this city, and her grandmother had just handed her a sweet. It was a dried date shaped like an Easter egg.

"Thank you, Grandmother. Can I eat it now?" she asked.

Yara's mother leaned over. "No, Yara. One more bite of your lunch, and then you can eat it."

Yara picked up her fork and poked at the salted cod on her plate. Tearing off the tiniest sliver, she looked up at her mother. "This bite?"

"No, Yara. A real-sized bite. Bigger, please."

Yara sighed and poked around at her fish. Finally, she picked up a bigger bite and shoved it into her mouth. Her mouth was full of partially chewed fish, and she barely managed to say, "I ate it."

"Okay, Yara. Go outside and play."

Yara grabbed a chunk of pé de moleque made with roasted garbanzos and a sticky artificial sweetener. She raced out towards the street where the other children were playing. Under the lamplights, she carefully nibbled on the crunchy treat while staring down at the chalk drawings of Jesus, flowers, and bunnies.

A gentle breeze touched her face. First warm, and then cooler. As she chewed, she noticed a new smell. Something she had never smelled before—a very faint scent of rubber, like a new bicycle tire.

She looked at her family and shouted, "Mama, what's that smell?"

Her mother didn't answer. She was talking to her uncle, and they were both laughing.

Yara looked down at her pé de moleque and took another small bite, snapping off a big, crunchy garbanzo. Small marks were forming on the chalk pictures. Scattered, irregular. They smudged the chalk in little circles.

They grew in number. She could see the marks exploding on the chalk drawings all down the street. The laughter had stopped. She looked up at her mother. Not knowing if she should be scared yet. She could feel that she was about to cry.

Her mother strode towards her, but she didn't look at Yara. She was peering up into the sky. Everyone was coming out into the street, facing the sky. But they didn't look worried. Their eyes were full of hope.

The rain fell down, covering the street. The many colors of chalk—pink, blue, and orange—blurred together in ghostly swirls that ran towards the sidewalk. Her mother was hugging her grandmother. They were crying.

Yara looked up at the sky and opened her mouth, letting the cool raindrops touch her face. Never had she seen rain, but she knew this was what it must be.

There was screaming, yelling, singing, and crying in every direction. She could hear people from far away dropping what they were eating, some half-dressed. At that moment, every single person was outside with their arms raised, reaching toward the clouds.

Chapter Sixty

The girls were at school, and Anela was at the night market. Makoa walked onto his patio and looked at the twinkling lights around him from the hundreds of patios in his termite tower. It was unsettling that the roof canopy obliterated the sky, and the only twinkling stars were directly in front of him. He could feel the dry, cooler air flowing downward, pushing the building's warmer air up through chimneys.

He stepped back inside and glanced down at his experience sphere. "Hey, Gene. Turn on *Look Up to God*."

The room dimmed as the show spread across the room around him. He sipped his cool tea and dropped down on the couch. An enormous raindrop landed slowly on a leaf. It bent under the bead of the water's weight before releasing the drop. At such magnification, the water looked viscous.

The music crescendoed at precisely the moment the leaf reversed its course. Some of the droplets that clung to the leaf ran back up when the leaf sprung backward. The experience was so vivid that his left hand began to rise. His mind yearned to touch the glistening liquid.

He leaned back into the couch, resting his cup on the armrest. The leaf faded to black, and Philippe came into view. His starched white shirt appeared especially crisp this morning.

"God is with you, my friend." Philippe opened his arms wide, his palms outstretched and his teeth beaming white. "I was thinking." He paused and looked up to the sky. It appeared he was searching for the perfect words. The sermon seemed to be coming to his mind at that very moment. "How is it that we lose our way? That we walk in the shadow, and that shadow is a weight upon our shoulders? It's because we are so easily distracted from the light. This is normal. We are human."

He stood and began to walk slowly. "We enjoy our electronics. We have work to do. We race home to feed our family and talk to our friends. We are busy creatures, always searching." He stared up into the camera. "'He that follows me shall not walk in darkness, but shall have the light of the life' (John 8:12). You may feel that you are walking in the shadows. You may feel that sometimes you walk alone. But you are not alone. That shade is not on you." His voice rose. "God is casting that shadow. God is looking over your shoulder at every moment. He protects you, always.

"'The people that walked in darkness have seen a great light: they that dwell in the land of the shadow of death, upon them hath the light shined' (Isaiah 9:12). You can see this light around you. Sure, today it may be a light you turn on with a wave of your hand." He chuckled, amused with his joke. "Or maybe a command to your AugWatch.

"But in the Bible, we speak of Heaven's oil. Remember the story of the ten virgins? Go ahead, look it up today (Mathew 25:6-13). The five wise virgins had prepared in advance for the darkness. 'Thine handmaid hath not anything in the house, save a pot of oil' (II Kings 4:2). They filled their lamps with oil. And what was that oil? It was the Holy Spirit."

He stopped, placed the Bible down and laid his hand on it. "Like many stories in the Bible, there is more than one message here. The wise virgins were bathed in light. They were prepared for the Lord. They had him in their hearts spiritually. But they were also prepared for the night. Well, in advance of the darkness, they filled their lamps with oil. My friends, embrace the bounty of God. Above you, clean water rains down. Below you, there is oil in abundance. The engines of progress were built on the oil God gave us from the ground. All of us are feeling the pain of our neighbors in Saudi Arabia. We beseeched them, and they are suffering for our sins."

Makoa leaned forward. He said out loud, "Did you just say we need to start using more oil?"

Philippe continued, his head slightly tilted to the side. "When God provides us with all we need, his blessings never run dry. Petroleum is used in the plastic for your serving bowls, the fibers in your clothing, shampoo, perfume, and even medical devices like hearing aids and heart valves. We use oil to ship goods." He straightened his neck and stared intently into the camera. "Water, around the world."

He closed his eyes. He breathed in through his nostrils and then opened his eyes with the exhale. "When we lose our way, it is time to stop . . . and to reflect. There is no drought from oil. It is a man-made event that has led to the collapse of Saudi Arabia—the wild fluctuations in the price of petrol will be hardest for our poorest citizens of the world. I want you to . . ."

Makoa shouted at the wall, "Gene, turn it off!" He stood up and stalked over to the window. His mind raced. Was he overthinking this sermon? Would Philippe really want to destroy Kasemchai's business?

Chapter Sixty-One

ANELA ENTERED the house and placed her bags on the kitchen counter. She called out into the living room, "Makoa! Are you there? I need help with the groceries."

She hefted the bags onto the counter. The heavy jars on the bottom were loud when they hit the surface and echoed around the quiet house. Walking into the family room, she saw Makoa staring out the window into the courtyard.

"Hey, can you help me unload the groceries?"

Makoa turned. His eyes looked glossy. They slowly focussed on his wife.

"What's wrong?" she asked.

"It's going wrong. It's all going wrong again."

"What do you mean?"

"I was just listening to the ministry. Philippe had this crazy sermon about oil. He said we need to bring back oil." He clenched his fist and stared at it, his lips pursed. He dropped his fist and said through gritted teeth, "This is just a game to them."

"Who's *them*?" asked Anela.

"All the rich and powerful. Philippe says we can only use water from the sky. Preap says water is water. Kasemchai starts making electric ships, and they all start running their ships on batteries." He was now shouting. "Do you know what happened?"

"Honey, calm down."

He shouted, "I'm not calming down! Do you know what happens? They convert all the ships, and all the people like me lose our jobs. First, the winery. Then, the fuel barges. I followed your advice and got to training for electric ships. And you know what happens?"

Makoa stormed over to the wall and slammed it with his palm. "Philippe wants all the electric ships to run on fuel again. Sure, it will be terrible for Sol-vang, but Kasemchai'll get over it. He'll be sitting in his skyscraper, crying into his cocktail. They don't care. This is all just a game to them."

He strutted over to the door and stared out at the backyard again. He threaded his fingers through his hair and grabbed his skull. Slowly, his elbow lowered to cover his face.

Anela spoke with a gentle but firm voice, pulling him from his dark thoughts. "Honey."

"What?"

"Your companion."

Makoa dropped his left arm and stared at his AugWatch. He jabbed at it with his finger. "What do you want?" he barked.

Kasemchai's avatar appeared in Makoa's cramped living room. "Sorry, did I catch you at a bad time?"

Makoa shut his eyes tightly and breathed in deeply. "No." He exhaled before his lungs burst. "Everything's fine. What did you want?"

"This rain in Brazil is crazy, isn't it?"

"It is."

"I just wanted to confirm you're coming over on Thursday."

"Yeah, I'll be there." Then he muttered, "Same time as usual."

"Great! Great. I'll see you Thursday."

"Sounds good." Makoa started to reach over with his finger to end the call when he stopped. "Hey, Kasemchai."

"Yeah."

"I heard Philippe talking today. Be careful. I think he may betray you."

There was a long pause. Kasemchai was quieter when he asked, "Why do you think that?"

"His sermon was something about oil and water being made by God."

"That's an odd combination," Kasemchai said. His avatar tilted its head just a little to the side. "And that made you think he'd betray me?"

"It's more than that. He said there's no shortage of oil and the engines of progress." Makoa waved his hand dismissively. "I don't know. I'm just telling you to be careful."

There was another long pause before Kasemchai answered. "Thank you, Makoa. You're a good judge of people. I believe you. I just don't know what this means. On the other hand"—his avatar laughed—"we're talking about the Labonne family. Of course they'll betray me."

Chapter Sixty-Two

ALESSANDRA'S HEELS made a quick staccato sound that echoed off the tiles as she hurried down the hall of the Palácio da Alvorada. She opened the door of the president's office without knocking.

The president was sitting in one of the large white loungers, staring out the large floor-to-ceiling windows. The distinctive concrete sail-shaped patterns on the outside of the building swooped down from their points on the left and right corners. There had been a rectangular pond and an expansive lawn many years ago. The pond was now dry, and the grass covered with dry gravel.

"President Batista, I have the proposal from the Seng family."

The president jumped up immediately upon hearing her voice. "Well, how is it?" asked the president.

"Interesting." She clasped her hands in front of her waist. "They're sharing routes with the Labonne family up and down the coast from Argentina to Canada. And they will fund the development of a canal toward Panama through what's left of Columbia. They say that once the canal opens, it will move far more water than the ships ever could."

"What do you make of this?"

"It's a serious investment they're proposing. In exchange, they want to lock up distribution rights between the two water families. And they will give us all the rights to the canal when it's finished."

The president paced back and forth. "We have to move the water if we're to have any type of economic recovery. We don't have any ships we can use for shipping water yet, and this pipeline sounds like the perfect solution to eventually take ownership of our distribution." He gazed out the window at a statue of two women in the empty pond. "But something doesn't sound right. Dammit, water is hard."

Carina, President Batista's digital companion, spoke up. "You mean 'strong,' Renato."

"No, it's hard. It's hard to wrap my head around, hard to manage, hard to understand."

"No, rocks are hard. Math is hard. Things that hurt your head are hard. Water is strong. Water can withstand great pressure. In the end, no matter how we try to exert our force on it, it always goes where it wants to go."

"Carina, are you philosophizing?" asked the president.

"I'm trying to illustrate my point, Renato." President Batista walked absentmindedly through the room as he listened to Carina's disembodied voice. "All water flows south. Building a pipeline that runs northwest through Colombia, across the Andes mountain range, would take decades to complete. The high altitudes will require constant adjustments to ensure the pipeline's integrity and efficiency. It might involve tunneling through mountains, which adds another layer of engineering complexity and cost. The geopolitical and environmental challenges . . ."

"Okay, okay, I get it, Carina. The pipeline is a red herring. But why would both regions want to share this new wealth? They should be more competitive."

"I was wondering about that myself," Alessandra said. "Brazil is significantly closer to the Labonne family. There's certainly enough water for both to share, but why join forces?"

President Batista's eyes came into focus as he realized his course of action. "Let's find out. Schedule a meeting with both families. And I want Liv Anselm to be there as well."

"Liv? From Sol-vang?" asked Alessandra.

"Yes, Liv. The Brazil-Sol-vang agreement was already set to make Recife a major shipping port. Liv can do more to help us ensure we don't ship all our water on the Seng and Labonne ships. I want a massive fleet of Brazilian ships as well."

"But the agreement was only to make military vehicles and retrofit tanker ships?"

"That was before the rain."

Chapter Sixty-Three

MAKOA STRODE down the hall towards the great living room. Usually, when he entered the room, it was well-lit. Tonight, he could see the experience sphere was on, but the sound was turned off.

As he walked closer, he could see it was showing the news. It was all about the storms in Brazil these days—the flooding. Kasemchai was on the couch. He had a blanket wrapped around his shoulders and his feet tucked up under it.

Makoa approached him. He stood at the far end of the couch, leaned over, and placed his hand on the couch arm. "You okay, boss?"

There was a long silence. Then a muffled, "I'm fine."

"You want me to turn some lights on?"

Kasemchai remained still for a long time, then jumped up, tossed the blanket onto the couch, and moved to the windows overlooking Central Park.

"Chanthavy, turn on the lights, not too bright," he called out. He stood in front of the windows, rubbing his face vigorously with the palms of his hands. "Yeah, I'm fine. I'm fine. You want a drink?"

"Sure, I'll have a booch. Whatever you got." Makoa stayed where he was, safely behind the arm of the couch.

Kasemchai wiped his eyes. He sniffed deeply, clearing his nose. He turned to face Makoa. "I'm sorry." He stared at the bar as if willing it to bring them a drink. "I've just been watching the rain in Brazil. It's just like Cambodia when I was a child."

"Yeah, I guess that's rough." Makoa thought about saying something, changed his mind, then changed it again. "When we were kids, we didn't like watching the news about Cambodia or the Ivory Coast. My parents always turned to another channel. My grandmother would mutter, 'All that rain and none of it here.'"

"I know what you mean. We never get the right amount of weather anymore. It's all or none." Kasemchai walked over to the bar. He pulled out a sachet of kombucha. "Here you go."

Makoa decided Kasemchai wasn't dangerous, just depressed. He walked up to Kasemchai and took the sachet, twisted off the lid, raised it in salute, and took a long drink.

Kasemchai returned to the bar and added ice to a silver martini shaker. He poured four ounces of gin and a dash of extra dry vermouth into the shaker, added the top, and vigorously shook it over his shoulder.

He poured the martini through the strainer into a tall martini glass, speared two pickled pearl onions on a long toothpick, and dropped them into the glass. He rested one hand on the bar and turned to face Makoa but didn't lift his head.

Finally, he took a long drink. "I don't know what's happening. I survived the storms. I'm stronger than this."

"Sure, man. You survived," Makoa said. "But that doesn't mean you're ready to relive it again."

Kasemchai took another long sip. "I remember a lot of bodies. They floated down the muddy water. I wondered where they kept coming from. Surely, the river would run out of bodies, but it never did." He looked out towards the window. "At first, they were bloated, filled with air like a balloon. Later, not so much. It was like the river was going through houses, pulling the dead out from the homes and sending them to the sea."

Makoa took a sip of his kombucha. "On the island, the water slowly ran out. People would fight. Then, they were too weak to fight. They would just lie down and die. We tried to drag them to the sea, but they kept washing back on the beach. Finally, we had to burn the bodies."

Kasemchai looked at Makoa. "I remember a sound."

They both smiled, and in unison, they shouted, "Flies!"

Kasemchai added, "So many fuck'n flies."

Makoa raised his kombucha. "No one needs to re-live this, man. You don't. I don't." He beheld the scene from the experience sphere. "The Brazilians don't need this."

"Yeah," Kasemchai said half-heartedly. He walked around in a lazy circle, stopped in front of the window, and stared out for a long time. "It's not the death."

Makoa thought his conversation had cheered Kasemchai up. But now he realized that was too simple. Kasemchai had moved on outwardly, but something was not right.

"It's not the death?" Makoa repeated.

"It's not the death and destruction that bothers me. I've seen all the death. I can handle more. No . . ." Kasemchai paused and closed his eyes. He inhaled deeply, and exhaled. "Sorry, you don't want to hear about this."

"What are you talking about? What could be worse than death and destruction?"

Kasemchai let out a long sigh. He shivered, and a little gin sloshed from his glass. He steadied himself and took another long sip.

Makoa lowered his tone. "Come on, man. I won't think you're an asshole. Or, at least, more of an asshole than you already are." He laughed, then stopped short. Had he gone too far?

Kasemchai stared at him. "Do you know why I've done all this shit? Why I invested in Sol-vang? The truth is, it wasn't to change the world." He returned to the couch, placed his glass on the coffee table, and slumped into the oversized cushions. "I wanted to show my father and Philippe that they could trust me to negotiate with the Saudis. I wanted them to see me as an equal."

Makoa thought about this. "That's okay. So you didn't want to save the world. At least you didn't invest in oil. You invested in batteries. They have to respect that."

Kasemchai leaned his head on the back of the couch and guffawed toward the ceiling. Then he jumped up. "Don't you see? Saudi Arabia is gone! I'll never be able to negotiate with the Saudis because the fucking country is done! There's no more oil coming from them. And now so many ships are running on my goddamn batteries that we probably don't even need the oil anymore. They'll just convert all the ships to batteries."

Now Makoa laughed.

Kasemchai stared at him like he had just gone insane. "What's so goddamn funny? I'm more of an asshole than you ever imagined. I don't care about people. I only care about a stupid meeting with the Saudis."

Makoa laughed harder. "I'm sorry, man." He laughed and laughed and then wiped the tears from his face with the back of his hand. "You're so focused on one thing that you didn't even realize that you've done something even more important."

"What do you mean?"

"All you ever wanted to do was negotiate with the Saudis. This whole crazy thing, all of the people you work with, just for one thing."

Kasemchai stepped up to Makoa and glowered at him. "I get it. I'm a horrible person. Why do you find this so damn funny?"

Makoa flung his arms open, palms skyward. "You said it yourself, Kasemchai. You didn't do this to negotiate with the Saudis. You wanted your father and Philippe to see how capable you are. To respect you."

Kasemchai considered this. "I guess so. But now what? My father wants me to work for the company—some kind of made-up position, Technical Visionist—and to walk away from Sol-vang. I showed them that I could do something as important as them, and then I ran back and went to work for them. Is that progress?"

"Well, A, asshole, you make it sound like having an awesome job in the largest company in the world is a compromise. And two—"

"B," Kasemchai corrected.

"What?"

"You started your example with A and then said 'two.' You need to continue with B."

Makoa stared with his mouth open. "As I was saying—and B, asshole—you always make every problem into a black-and-white choice."

"What is my other choice?" asked Kasemchai.

"I don't know yet. But when I was a kid and had to make a difficult decision, my mother would pick up three stones." He walked over to the table and picked up three coasters from a stack. "Or three of anything."

He threw the first coaster on the table. "She would say, 'The first stone represents the logical decision your mind wants to make. The decision everyone believes you'll make.'" He threw down the second coaster, a foot from the first one. "The second stone is what your heart wants to do." Then he threw the third coaster down in the middle. "The third stone isn't the middle. It's not a compromise. It's the infinite universe of possibilities. You must listen to the third stone to find the correct path."

Kasemchai stared a long time at the three coasters. Makoa leaned back into the couch and said nothing.

"You know . . ." Kasemchai put his finger on the middle coaster and moved it around the table in little circles. "Your mother was a smart woman."

"She was."

"Okay. So think about this, Makoa. A, we're the second largest water family, not the first. And B, see what I did there?"

"Yeah, yeah. Get to your point."

Kasemchai grinned. "And B, he didn't actually say I had to walk away from Sol-vang. He said something like I need to free up time. In fact, I don't need to be like him. I can still do something both he and Philippe haven't done."

Makoa raised his kombucha. "That's right, man. The infinite world of possibilities."

They both took sips from their drinks and sank back into the cushions.

"So, how did your family become so successful?" Makoa wondered. "I know you said your father was in construction, but how did it actually happen?"

Kasemchai stared out the window for a moment, collecting his thoughts, and then he began, "Well, his company wasn't anything fancy, but he had a lot of deals in Cambodia, Laos, and Thailand. When the rain started, we were in Siem Reap. We just barely made it out alive. We drove straight through Thailand to Laos."

He sat up and took another sip of his Gibson, staring into the icy liquid in his glass. "We didn't stop for anything. My father just wanted to get as far from the rain as possible. We finally reached Vientiane, in Laos. In the early days, the Chinese started building waterways to collect water, but they had to go through Laos and Thailand. My father recognized what was happening and started making deals with everyone. He built canals and pipelines to direct the water."

"How did he know where to make the canals?"

"He didn't. He started just outside of the rain, but it kept expanding." Kasemchai leaned forward and rested his elbows on his knees, his fingers laced together. "So, he kept rebuilding the canals and pipelines. In the early days, it was chaos. People were just trying to live. My father gave people money to work on projects."

"I guess he was just in the right place at the right time," Makoa said.

Kasemchai shook his head. "No. That's not it at all—lots of people were there at that time. But he did something about it. He made an empire from the water that fell from the sky." Kasemchai sat back on the couch. "That's my father. If I think about it, I can tell why my father and Philippe are so close. They pretend to be so different in public, but they need each other. There's no one in the world who's done what they've done. That's why I'll never be exactly like them."

"What do you mean?" asked Makoa. "You've built an empire of electric ships." He sniggered. "You've put thousands of tankermen out of business." Then he dropped his gaze. "Sorry, too soon."

"Sure, it's easy to start a business when you begin with lots of money," Kasemchai said. "And besides, the real genius is Liv."

"Well, whatever it is. You made a difference for me. I mean, this job of hanging out with you is fine, but the Sol-vang Reskilling Program was genuine. I have gone from failure to failure. I really felt that, for the first time, I was doing something that I could feel good about. That working on electric ships was the beginning, not the end."

"Thanks, Makoa. I guess that helps." He sat back and smacked the arm of the couch. "And good timing. I'm flying to Sihanoukville next week to meet with my father."

"Good. Good." He stared intently at his feet. "There's just one thing, Kasemchai."

"What's that?"

"If the Labonnes back away from Sol-vang, go back to oil . . ." He took a sip from his sachet and stared at the condensation forming on the outside. "I lose again. People like me lose again."

"What do you mean? That's why I created the Sol-vang Reskilling Program."

"I just want the change to stop. Whatever I do, I'm good at it. I just want to work. I just want to know my family is safe."

Kasemchai carefully placed his glass on the table in front of him. He leaned forward, rested his elbows on his knees, and clasped his hands together. "No one is safe. When I meet with my father and the Labonnes, they will try to hurt me. That's what these people do. They will betray me. I have betrayed them."

He sat up and swept his arm across the New York skyline outside the windows. "There was never a world where you could just do your job. There is only a world where you reinvent yourself every chance you can. You get better, stronger."

"Shit. Well, have fun."

Chapter Sixty-Four

Kasemchai had only been in Sihanoukville two days before he found himself in the dragon room, seated across from Philippe and Angélique. The room was impressive, with tall, ornate lamps casting a soft upward light that reflected off the intricate ceiling details. The light then cascaded down the columns with bulbous bases, giving the impression that gold was melting onto the floor.

On the wall behind Preap hung his most prized possession, a gift from the President of the People's Republic of China: an enormous artwork of a black and gold water dragon breaking through a turbulent sea, with sapphire and green jade streaks. However, it was not a painting; it was made up of millions of tiny flakes of different colors that were meticulously attached to the canvas piece by piece. These flakes shimmered with the slightest movement in the room, and the iridescent colors swam across the canvas. The eyes of the dragon seemed to follow the guests, and many people claimed that the dragon moved when they turned their back.

Preap lifted a carafe of wine and filled the four glasses on the low table before him. He then slid them across the table toward Philippe and Angélique. As he pushed the glass toward Kasemchai, he said, "I don't have to tell you how hard the floods were when the Rain Wall appeared. But through sheer luck . . ." He lifted his glass toward Philippe.

"Or through divine intervention," Philippe said.

"Thank you. Whatever your belief, our two families pushed through tragedy and found a way to harness the water and ship it around the world. It is our ethical imperative to protect our most precious resource."

"It wasn't easy, and your father and I didn't always see eye to eye," added Philippe. This brought a smile to his face, and he turned to Preap, who returned the smile. A thousand shared memories flashed through their minds. "But we forged an alliance that allowed us to make the best choices for all of humanity. We have streamlined our decision-making process to help ensure we can ship the most water possible at the lowest cost possible."

"That's right, Philippe. And we are shipping water to more places every day. There is no shortage of water. The real challenge is preparing a port to receive water and distributing it from that point of entry."

Philippe laughed. "It's our job to supply water to everyone with minimal drama, and when people believe there are only two choices, we always win. Do you remember the riots when Preap started shipping water to South Africa?"

Kasemchai turned to Angélique, who shrugged, unsure where her father was going with this story. He turned back to Philippe. "Sure. I always thought it was a little strange that we shipped water to South Africa when your ships were so much closer."

"That's right," Philippe said, raising his eyebrows. He was clearly enjoying the story. "Much closer. What you didn't know was that President Ndlovu was trying to encourage Angola to ship water to South Africa. We opened a route for Preap's ships and launched a campaign to promote the health benefits of Asian water. People went berserk."

"Sorry, why did people go berserk?" asked Kasemchai.

"I don't want to sound flippant, but the water from Africa is from God. It didn't take long for people to pick sides on which benefits they preferred: the water from the Labonne family or the water from the Seng family."

"But that's absurd. It's all water."

Preap shook his finger in the air. "That's correct."

"People like binary choices. We . . ." Philippe pointed at Preap with his two fingers and then to himself. "Provide choice. They demanded our water over Preap's water. We struck a deal and declared victory over Asian water."

Preap and Philippe leaned forward and tapped their glasses together.

"We removed the Angola connection," Preap said.

The two men threw their heads back and laughed heartily. Angélique rolled her eyes.

"That's the reason we brought you in, Kasemchai. You and Angélique couldn't be more different than Preap and me. Regardless, there is one thing that unites us all: the two families. Our strength is the families. Our wealth is the water."

Philippe leaned back, and the two men exchanged subtle glances, a wordless agreement passing between them.

Angélique leaned in toward Kasemchai. "Brazil is on the brink of recovery from years of drought, and soon they'll recover from the flood. We can't let them capitalize on this newfound wealth before we do."

Philippe tapped his hand lightly on Angélique's arm. "The rain is a national resource that clearly belongs to Brazil. We must be cautious in our language, Angélique."

She immediately averted her gaze after this rebuke. When she looked back up at Kasemchai, she looked defiant, daring him to say something.

Philippe continued, "We cannot swoop in and take ownership of these water rights. What we can do is help this country capitalize on this newfound resource as effectively as possible. They do not need to repeat the mistakes and struggles Preap and I endured."

"Kasemchai, what the Labonnes say is true," Preap said. "We should establish partnerships with the Brazilian government and key stakeholders. Let's offer aid, invest in infrastructure, and fund recovery projects. In exchange, we secure the rights to ship the water."

Kasemchai looked back and forth, then leaned forward. "What about the other countries in South America? Why only Brazil?"

The water dragon on the wall watched Kasemchai as Preap spoke.

"The rain covers an area across the northern tip of South America. The countries north of Brazil are so flooded that there's no government or infrastructure to work with. The only country that can manage the water is Brazil."

Kasemchai thought about this. He pictured in his head what the countries looked like from the maps he'd seen on the news. The rain cut through most of Venezuela. Trinidad and Tobago, Guyana, Suriname, and French Guiana were all gone. This Rain Wall was estimated to be the size of the combined African and Asian walls.

"Why are you telling me all of this now?" asked Kasemchai. "I'm the Chief Technology Visionist. I don't see what technical solution you're looking for."

Philippe didn't wait for Preap to answer. "Because, Kasemchai, we must ensure that the complex infrastructure and shipping remain solely under our control. This is where our true power lies. The country may own the water, but we must control how it's accessed, stored, and distributed."

Kasemchai's face was stoic, but inside his mind, he was frantically trying to put together the puzzle. His father could have called him to have this conversation. What was in Brazil that only he could solve? He struggled with the thought until it finally dawned on him.

They needed something from him. It sounded delicious saying it in his head.

"Thank you all for bringing me into your planning. I . . ." He leaned in to show his confidence. "How can I help you?"

Preap looked at Philippe and then back to Kasemchai. "Philippe and I have discussed the risks to our empire many times. The greatest one of all has always been, quite frankly, the Earth correcting itself."

"Huh, correcting itself?" asked Kasemchai.

"Yes, the earth is correcting itself. It's not natural or predictable that all the rain in the world would fall in two places. The heat and carbon pollution were too extreme, and it was only a matter of time before the Earth would respond in an erratic way."

Kasemchai regarded Philippe. He wondered if he had an answer to how God had abandoned humans. "Isn't this just God's way?"

"I cannot speak for God, Kasemchai." Philippe sat back and placed his hands in a steeple. "But in the book of Genesis, it says that God placed Adam in the Garden of Eden to 'work it and keep it.' It was our responsibility to care for the Earth. We did not do well. I believe the great flood wiped away man's sins, and then the great drought burned away man's mess."

Angélique slid her finger beneath her choker, playing with it absent-mindedly as she spoke. "Our families were rewarded for protecting the water. We shared this precious resource with the world, and our tanker ships covered the ocean. But then, something changed."

Kasemchai scanned Preap's face and then stared at Angélique in astonishment. The source of his power over them was ringing in his head. It was the electric ships. "Are you saying you think Sol-vang has caused the Rain Wall in Brazil?"

Philippe held up his hand. "No, I'm not saying that converting ships from oil to batteries reduced greenhouse gasses enough to change the environment. But it certainly contributed. It may have even accelerated the change. And change means water in places like Brazil."

Preap frowned sternly at his son. "I told you that electric ships could destroy our family business. But it's more complicated than that. She's outmaneuvered all of us."

"Who?" asked Kasemchai.

"Liv." Angélique stabbed at the table with her index finger. "Sol-vang is literally powering a huge number of our ships. The Brazil-Sol-vang agreement will create a network of ships that can go up and down the coast, on both sides of the continents."

"How?"

"Simple. Brazil acquires Columbia or Venezuela and ships right over Panama. With Liv's help, Brazil will completely lock up two continents."

"The families must move in the same direction," Philippe said. "We want you to choose your loyalty."

Preap sat back and smiled warmly at his son. "I have already spoken to our attorneys. We're drafting papers now to present to Liv. We wish to have all of our engines returned to our ships."

Kasemchai slumped into his chair. "This could be devastating to Sol-vang." He looked up at Philippe, his eyes betraying his concern. "And what about you, Philippe?"

"I would also put the engines back in if I could, but Angélique and I did not have the foresight your father had. We may take these ships offline, but for the moment, we won't make any changes. We have already begun our campaign to increase the use of oil around the world." Philippe placed his elbow on his armrest. He leaned in toward Kasemchai and gestured with his free hand. "More importantly, we must ensure that Liv doesn't attend our negotiations with Brazil."

Kasemchai was getting hit on multiple fronts, and the conversation was spiraling out of control. His father's desire to return the engines would be a huge blow to Sol-vang's finances and brand. The idea that his ships had reduced carbon so much that it caused rain to fall in Brazil was wild speculation, but what if it was true? Was it an existential threat to the family empire? And how would he tell Liv not to attend the meetings?

He took another deep breath. "You may be correct. It's possible that Sol-vang reduced carbon emissions so much that it caused the rain in Brazil. But who's to say that if we continue to pollute the earth, the rain won't stop entirely in every region?"

Angélique leaned forward in her chair. "And who's to say that Sol-vang won't lead to more unintended consequences? Liv is developing military contracts with many nations. She says she's trying to help humanity. Military contracts don't sound humanitarian, and think of all the people in the oil industry she's put out of business."

Kasemchai snapped back, "I am personally running the Sol-vang Reskilling Program to help people by removing them from an unhealthy environment and giving them the skills to support a cleaner, more sustainable infrastructure."

"We know about your conversations with the Hawaiian," Angélique said.

He had no idea what she meant by that, but he was certain she had intended to demean him. "My conversations with the Hawaiian? What does that mean?"

"Are we no longer good enough for you? Do you need to spend your time with people like your Hawaiian friend? One of your cars is worth more than he will make in his lifetime."

The veins on Kasemchai's neck bulged as he growled, "Has wealth changed you so much that you can no longer recognize that the decisions we make affect millions of people? I'm just trying to find out what it's like to be on the receiving side."

Angélique could only open her mouth before Philippe jumped in. "Woah, woah. Hold on a minute, Angélique."

"That's right," Preap said. "I love the passion you share for our business, but let's get back to the purpose of this meeting." He took a quick sip of his wine. "Kasemchai, we have been discussing something else. We would like you to lead the negotiation with Brazil."

Kasemchai's head snapped to his father. "What? You want me to lead the negotiations?"

"That's correct. Philippe and I built our empires with brute force. We need someone with a lighter touch who can help the Brazilians recognize the value of our partnership."

Kasemchai stared down at the table. "Why me? Angélique has been closer to your business these past few years than I have."

"That's a good question," Angélique said as she pulled at her choker, letting it snap against her neck. "Do you remember the last time we were here? What was that, like ten years ago?"

"Sure, nine, ten years ago."

"Do you remember how hard that was?"

Kasemchai stared at her. Was she going to tell her father about their hookup? He looked towards the fathers but didn't dare stare into their eyes. His heart rate increased. He tried to slow his breathing to calm his voice.

"I do."

Angélique smiled broadly. She was enjoying this. She waited until the silence was uncomfortable, then said, "You told me you wanted to do something important. This is your opportunity to show our families where your true loyalty lies."

Kasemchai's heart rate slowed, but his heart still beat so powerfully he was certain the others could hear. He looked up at Philippe. "And you think the Brazilians will trust me more."

Philippe chuckled. "I'm glad you recognize that, Kasemchai. That's exactly why we want you to lead the negotiations. We know the Brazilians have already reached out to Liv. President Batista will listen to you."

Preap leaned in. "We need you to make sure your number one priority is the water families. And don't let Liv anywhere near our negotiations."

Chapter Sixty-Five

KASEMCHAI STOOD in his office in front of the enormous windows, staring into the rain. He caught an individual drop and watched it fall until he couldn't distinguish it from the others. He couldn't see the rain hit the ground, but he could see the ground was wet. He thought how strange it was that something happening right before his very eyes was invisible to him.

Chanthavy interrupted him. "Liv is calling, Kasemchai. Would you like to speak with her?"

He didn't turn. The ground was nothing but a blur. "Yes, please."

Liv's avatar appeared, standing behind Kasemchai. He blinked and stared at his reflection in the glass. Liv's avatar was also in the reflection, standing behind him. Its hands were clasped together in front of its waist. The eyes looked concerned. He turned around and walked towards the chairs in his seating area.

"What's up, Liv?"

Liv's avatar walked with Kasemchai toward the new lounge chairs, which were covered in crushed plum-colored fabric, low-backed and featuring large angular arms. He sat in one of the chairs while Liv's avatar sat across from him.

"I need to tell you something," she said. "There's been a change to the Brazil Sol-vang project."

"Okay." He said it slowly and an uncomfortable grin came to his face. "You're being strange. Usually, you just start talking, Liv."

"Will you be attending the negotiations with Brazil?"

"I will. In fact, I will be representing the Seng family."

"Really?"

He stared at her avatar. Was she surprised he was negotiating or surprised he'd be there? "I am officially on the payroll now. Remember, Chief Technology Visionist. In fact, we just recently had a meeting about Brazil."

For days, he'd been imagining this conversation. He had a perfect setup, but the next words were not how he wanted to open this conversation. "I heard something strange. My father and Philippe believe that Sol-vang caused the rain in Brazil."

He watched carefully to see how her avatar would respond. While an avatar couldn't display the true emotions of the person they represented, its actions often reflected those emotions. If she looked down, her avatar would look down too. But Liv's avatar only stared directly into his face.

"I see," Liv said. "Competition. Rain has made Philippe and Preap very powerful, and they don't want to share that power with a third party."

"It's not about competition, Liv. That's what I spoke to them about. They want to help Brazil recover. They've been there, and they can help Brazil avoid the mistakes they made capturing water."

Liv's avatar turned its head to the side. It sighed and continued, "President Batista isn't an idiot, Kasemchai. He knows they want to control his water. It's true they're very poor, and their infrastructure is in shambles. But they have something, and they know the wolves are circling."

Kasemchai stood up and began to pace. "Be careful, Liv. You don't want to side against the two water families. And don't forget, I belong to one of the water families."

Liv's avatar's brow furrowed as Liv mulled over her response. She finally said, "I hear what you're saying, Kasemchai."

He was being too intense and realized Liv was talking to him like he was an idiot. He forced a casual laugh. "I'm sorry, Liv. I'm being very dramatic. What is your big change?"

"Well, as you know, we were already planning to build a new complex in Recife to produce the battery-powered vehicles. That also included an agreement to let us improve the port and build a new EC there."

"That's great. Now, more ships can dock in Recife and swap batteries. That will make Recife a huge shipping center."

Kasemchai could see Liv's avatar's back straighten as it embodied her enthusiasm. "Sure. And that's only the beginning. Just south of Recife is the Atlântico Sul Shipyard. It's the largest shipyard in South America. We're working on an agreement to invest there to make electric ships."

The grin on Kasemchai's face slowly faded, and he spoke in a serious tone. "You're quite connected in Brazil, Liv."

Liv didn't realize his tone had changed, and she continued with the same enthusiasm. "Serendipity, my friend. It was only luck that we had already begun our commitment to the government when the Rain Wall appeared. Between Recife and the Atlântico Sul Shipyard, we should be able to dramatically increase shipping as far north as Montreal and south to Buenos Aires."

Kasemchai placed one hand on the back of his chair, and the other hand on his hip. "With new investments in infrastructure to move the water to the coast and new ships, I suppose you think this will help the water families?"

The avatar's mouth opened and closed a few times, shaping a silent "O" as Liv searched for the right words. "Well, yes. And that's what I wanted to talk to you about. The Brazilian government has asked me to work with them in the negotiations."

"Do you mean that you will be their negotiator, Liv?"

Liv's words stumbled over his, "No. No. As a consultant."

"So let me be clear. You will be in their corner to make sure they can really stick it to my family. Does that sound correct?"

Liv's voice rose in pitch. "Not at all, Kasemchai. They know your father and Philippe are very powerful. They want to make sure they are not exploited. And to help ensure we reach an agreement that is best for all parties."

Kasemchai gripped the back of the chair more firmly, his knuckles burned white against his cinnamon-colored hands. "Liv, I am just settling into my role in the family business. Both families have entrusted me to lead the negotiations on their behalf, and now you bring me two problems. Sol-vang has quite possibly brought about the greatest threat to our family, and you, my business partner, will literally be sitting on the opposite side of me in the negotiations."

Liv's avatar stood up and began to pace. It walked back and forth between the chairs and table like a caged lion. "I am not your problem!" It stopped when Liv realized just how loud she shouted this out. "You and I have the same dream: to create a world where the largest ships not only carry goods but also bring the promise of a cleaner, more sustainable world. Maybe we did cause the rain to fall in Brazil. But Kasemchai, if we are responsible for bringing water to a whole new region, this is a gift. It's water, not a commodity to sell. It's what we need to live."

Kasemchai gripped the chair with both hands and leaned over the top. He remembered clinging to the roof in the flood when the rains first appeared. He could taste the vomit in his mouth as he relived that day. Was she using his father's words against him?

"That was your dream, Liv!" Kasemchai shouted. "It was always your dream. You're the battery girl. My dream was to lead the Seng family."

"Why do you have to choose between us?" Liv's voice trembled, raw with emotion. Her avatar pressed its hands tightly to its chest. "And this wasn't just my dream. It started with batteries. We made something that could help everyone—you and I."

Kasemchai's face was emotionless, and his voice deadly calm. "No, Liv. I'm an investor." He walked away towards the tall windows, then turned back abruptly. "I'm a Seng."

Chapter Sixty-Six

NEARLY A WEEK HAD PASSED, and Liv was still upset about her last conversation with Kasemchai. But that wasn't what bothered her at this moment. She looked up slowly when Chantal entered her office.

"You look tired, Liv. What's up?"

Liv hung her head over her crossed arms at her desk and then looked up again. Her face was bloodless. "I've just received notice from the Seng attorneys."

"Notice about what?"

"They want to return all the engines to their ships. It states the electric ship pilot is finished."

Chantal's mouth dropped. "Finished? What pilot? We've converted hundreds of ships."

Liv stared out the window with glazed eyes. "Eight hundred and seventy-two." Then, she was silent. With her hand, she made slow circles on her desk. Stopped and tapped her knuckle a few times. "Noa, ask Hendrik and Aksel to come to my office."

"Do you want me to leave?" asked Chantal.

"No, I want you in here also. We need a plan."

Chantal walked over to the chairs and sat down. "Did you really save all the engines?"

"Of course we did. They're stored in a field, mostly in China. The first fifteen outside Rotterdam."

"What's that going to cost us to put them back?"

Liv smiled. "That's just it. I agreed in the contract to save them. We never said we'd cover the cost to restore them."

Chantal's brilliantly white teeth appeared from the grin that sprang to her face. "Ah, Liv. You are a genius."

Aksel entered the office first. "What's up, boss?"

"I need to do an audit immediately of all the engines we're storing for Preap. I want to know the exact location and status of each engine."

"Why now?" he asked.

Hendrik entered the door at that moment. He leaned his tall body against the door frame. "Every engine for Preap?" he asked.

"That's right. They want every engine put back in place, and I need to know exactly what I'm working with."

Hendrik slapped his forehead and stared up at the ceiling.

Aksel spoke in a quiet voice. "I have all their details, Liv. We just did an audit two months ago. But . . ." He looked away. "I have bad news."

Liv jerked her head up when she heard this. An expression no one had ever seen in this office before crossed her face. Her eyes narrowed. It was anger—almost hatred—she leveled at Aksel. "If you have bad news, why am I just hearing about this now?"

Aksel started toward Liv's desk, then stopped himself. "I'm sorry, Liv. I wanted to tell you, but the damage was already done. You were so focused on Brazil. I couldn't get time to share this with you."

"Share what?"

Aksel flailed his hand. "It looks like, well, many of Preap's engines are covered in pitting. We sealed them properly, but it appears moisture has gotten into them. A lot of moisture."

"What does that mean? Are they ruined?"

"Well, maybe not ruined. But they're going to need extensive rebuilding."

Liv leaned back in her chair. A wounded sound like "Ahhhhhhh" escaped her lips.

"Are we really going to put the engines back?" asked Hendrik. "What do we tell customers when they hear one of our biggest customers has ended our agreement? There will be panic. People will rightly want to know what is wrong with our system or electric ships."

Liv sat up straight and slammed her hand on her desk. "Everyone, calm down. They're posturing. They want something. I don't know what yet, but this is just a move. It's chess." She stopped and looked at Aksel. "And we're not putting any engines back yet. I just wanted to know the state of our liability, and clearly, it's shit."

She dropped her head down on her arms and shut her eyes tightly. With a muffled voice that barely escaped her arms, she mumbled, "We need a plan. We need a strategy. We need . . ."

Liv practically jumped up from her seat when her AugWatch buzzed. She read the name of the caller and hissed, "Shh, no one say a word."

Liv tapped her AugWatch, and Angélique's avatar stood in the middle of the room. Her shiny bobbed hair flared wide over her shoulders. Her dress was tight and black, with stiff shoulder pads that stuck out like tiny wings.

"Hello, Liv. It's been ages."

Chantal visibly shuddered.

"Mrs. Camara, such a pleasure to see you," Liv said.

"It's so nice to see you too. I was thinking of you lately. I hear so much about the work Sol-vang is doing in Brazil. It's so important. They need all the help they can get." Her avatar walked up and sat on Liv's desk. "Don't you agree?"

Liv absentmindedly wiped imaginary dust off her desk. "It is indeed. The new factory in Recife will bring many new jobs to the country. Kasemchai and I were just talking about this. Did you hear from him?"

Angélique's avatar cocked its head and wiggled its eyebrow with a devilish flair. "Oh, you know that boy. If I want old news, I'll go to him. I have much better sources for my information."

"Ah. That's good, I guess."

"I guess so. These factories in Brazil, I understand that you will only build power units for non-lethal heavy-duty vehicles. Is that correct?"

A smile briefly graced Liv's lips. Angélique did have good sources. "That is correct."

"That's great. The world always needs more troop carriers. Are we alone, Liv?"

Angélique spun her head around and scanned the room. A wave of heat washed over Liv's face, and her neck bristled with fear. It took a moment to recover when she realized it was just a trick. Her avatar couldn't actually see anyone in the room. Somewhere around the globe, Angélique was in her own room, talking to an avatar of Liv. The reality was an illusion.

"You can speak in confidence," Liv said.

Angélique's avatar leaned in just a little, and her smile was wicked. "I'm aware that Preap has asked to have his engines returned to his ships. That must be very disconcerting to you."

Liv studied the face of Angélique. Her grin was surely intended to raise her blood pressure. "I see this as unfortunate for the world. This will surely lead to an increase in carbon pollution, and that can't be good for the Earth."

Angélique held her hand up and studied her nails. "I suppose so. But you know what won't go up? My service contract." She whipped her hand down like a knife and stared right into Liv. "That was very clever of you to tie our contract to the cost of fuel. With the chaos in Saudi Arabia and your very successful effort to convert ships to batteries around the world, the cost of oil is skyrocketing. You owe us this to help bring our contract fees back down."

Liv was stunned. It was all too complicated. Angélique could have just reached out to renegotiate. Was this really just an elaborate scheme to reduce her costs? She opened her mouth to speak when Angélique cut her off.

"I also understand you will be helping Brazil build new tankers. This is great because we need lots of new tankers. Well, it would be better if they ran on bunker fuel, but that's neither here nor there. What I wanted to say is that at this very moment, we are negotiating with Brazil to help them bring their water to the world. And by *we*, I mean the two water families."

Liv tried not to sound sarcastic. "I was aware of that."

"I thought you would be. I am also aware that you are consulting with President Batista. I'm so glad you are bringing your expertise. I have a suggestion for you."

Angélique was truly testing Liv's patience, but she managed to stay professional. "Please, share it with me."

"I need you to focus on the new plant you're opening in Recife, building those tankers we'd like to buy. And, starting today, I need you to stop providing any help with the negotiations between Brazil and the water families."

"I'm only a consultant when they need me, Angélique. I won't even be in the negotiations."

Angélique looked down and swatted at an invisible item on her bare knee. She admired her leg for a moment. "I don't think you heard me clearly. Nothing—absolutely no discussion whatsoever with the Brazilian government about water. If I hear anything to the contrary, you can consider our contract dead."

Liv stared straight through Angélique. She could see the others in the room but didn't even dare look at them for a second. "I understand you perfectly, Angélique. I will cease conversation immediately with President Batista's team."

"Not just his team, Liv. I mean everyone. You do not consult on anything but your troop carriers and the ships you're building. If we have to end our contract, then I will have to take ships offline. Do you know what that means?"

"You ship less water."

"You're good, Liv. That's right. We ship less water." Angélique's avatar stood and began to walk away, then stopped and turned back. "And Liv, people need water to live."

Chapter Sixty-Seven

KASEMCHAI TOOK breakfast in his suite overlooking the Pinheiros River in São Paulo. The hotel had been remodeled so his father, Philippe, and Angélique could stay there during the negotiations.

The setting sun made it harder to see the garbage and debris floating down the muddy river. It was swollen. Carving out a path through the ruins of buildings that used to sit on its bank, but it was filled with water —a miracle people had long stopped believing in. Just the same, it gave Kasemchai no joy in looking at brown water. This water only signaled hardship ahead.

"Chanthavy, call Liv," he called out.

She took an uncharacteristically long time to answer. When her avatar appeared, it sat at the table across from him, appearing to anxiously survey the room.

"Hey, Kasemchai. I don't think we should talk now. Is it quick?"

"I won't take long, Liv. I just wanted to check in with you. We didn't end our last call on the best of terms."

"It's okay. I forgive you."

He gave a crooked smile. "You forgive me?"

Liv's avatar played with its belt for the third time. "I'm sorry. It's just . . . I don't think we should talk right now."

"Look, I'm in São Paulo."

"I know."

Kasemchai stopped. He rolled his eyes as he processed this information. "You know? How do you know?"

"Shit. Look, I promised Angélique I wouldn't go near the negotiations."

Kasemchai thought about this a moment. They had agreed that he would tell Liv not to attend the negotiations. He had meant to tell Liv earlier than the day of the negotiations, but Angélique clearly didn't trust him to take care of the job.

"What did she tell you?"

Liv spoke without pausing. "I don't want to go into the details. But I think it's best to let you work directly with the president's team. I don't want to be anywhere near it. I'm focused on the Brazil Sol-vang project, and I need to make sure that it runs smoothly. I'm only here for the night, and then I'll catch a connecting flight to Recife." She took a deep breath and exhaled loudly.

Kasemchai said slowly, "You're in São Paulo?"

"Yes. But look, I know you want to lead the negotiations with them, and I don't want to interfere. You need to do what's best for your family."

"Okay, so it sounds like you won't be in our negotiations. That's good. That's what I wanted to call you about."

"Great, well, good luck. I'll talk to you in the future. Let me know how it turns out."

"Sure."

Something wasn't right. He thought it would be harder to tell her to stay away from the negotiations. "Liv, is this what you want?"

Her avatar looked at its hands for a long moment, rolling its hands around each other, clutching, releasing, rolling. It said nothing.

Kasemchai couldn't wait any longer. "I'll answer for you, then. You think I should let Brazil build its own shipping fleet, that my family's wealth is big enough and should stay in Asia."

The avatar's eyes looked watery, its eyebrows pinched inward, carving faint creases across its forehead. Very quietly, it said, "No, that's not it at all. Do you have to suffer for someone else to grow? Can't you both grow? Together? They're not taking your water. Brazil is bringing new water."

Kasemchai said nothing.

Liv's avatar gazed into his eyes and quietly asked, "Why did you join our dive trip in Ishigaki Jima, Kasemchai?"

"What do you mean?"

"You never told me. You just showed up like the captain knew you were coming. What really happened?"

Kasemchai stared at his hands for a moment. "Do you remember when we returned to Ishigaki Jima, the awamori distiller came to visit us?"

"Of course. That's when you decided to reconnect with your father."

He laughed. "That's true. What I didn't mention was that when I visited the distillery, I saw you walking down the street."

"What do you mean?"

Kasemchai gave a hesitant smile, then looked down. "I mean, I was a bit of a prick. I was going to buy his distillery . . . and he wasn't happy about it." Kasemchai raised his head, a twinkle in his eye. "And then I saw you. You turned and stared right at me."

"I did? I don't remember that."

Kasemchai snorted. "No. No, you didn't see me. You looked right through me. But I saw you, and I just knew I had to meet you. There was something about how you stared at the ocean. I wanted you to look at me like you looked at the sea."

"Why didn't you ever tell me that?"

"Because . . ." He thought about what the reason was. "I don't remember exactly. We just got so busy."

"I guess we did." Her avatar glanced over its shoulder. "We drank a lot that night. I'm kind of surprised nothing happened."

Kasemchai averted his gaze. When he looked up, a sheepish grin came across his face. "Haruka."

"What about Haruka?"

"Remember when she saved my life after the water snake? I had nearly run out of air, and she came to me. It was very intense. I was . . ."

He looked away and back again quickly. "I was conflicted. I came for you, and then she complicated things."

"Two good choices."

"And I chose neither."

"Hmm." Liv's avatar stared serenely into Kasemchai's eyes. "None is a choice, Kasemchai."

"No. None is not a choice. Maybe it was the right choice that day, but I could have made a choice every day after that." Kasemchai looked down at his breakfast plate and poked at a piece of fruit with his fork. "And yet I stayed with none. I guess I'm just a conventional guy."

She laughed. "Do you remember my toast that night?"

"I'm not sure I do."

"We are traitors to convention." Liv raised her hand. "Here's to courageous change."

Kasemchai raised his hand in an imaginary toast to her.

They stared at each other for a long moment. Then, quietly, Liv said, "I have to go, Kasemchai. And you need to do what your father wants—what the water families want."

Chapter Sixty-Eight

A GROUP FLANKED PRESIDENT BATISTA. Alessandra was just off to his right.

President Batista turned to her and asked again, "Will she be here?"

Alessandra leaned in and whispered into his ear, "No. She said she can't participate in the negotiations. The conflict of interest with Kasemchai would make it impossible for her to be objective."

President Batista turned and hissed, "I thought we already made it clear that she was only a technical advisor."

"She understands, sir."

The door opened at the far end of the room, and President Batista's attention snapped back in that direction. Philippe, Angélique, Preap, and Kasemchai entered the room, along with advisors and security teams for each family.

President Batista strode forward purposefully and threw his arms up into the air. "Welcome, my friends! Are you thirsty?"

Philippe, Angélique, Preap, and Kasemchai responded in near unison, "We are not thirsty."

"We are so honored to share this moment with the two great water families." He leaned forward and shook hands with Philippe, and then he and Angélique leaned in and kissed each other's two cheeks.

Preap reached out and shook the president's hand before stepping back slightly and turning toward his son. "I'd like to introduce you to my son, President Batista. Kasemchai Seng."

"My pleasure completely, Mr. Seng." President Batista reached out with his left hand, placed it over their two hands, and shook vigorously. "I'm very familiar with your company."

"I hope you mean Sol-vang and not my dating PrefX, sir."

President Batista threw his head back and laughed. "I had completely forgotten, Mr. Seng. You have changed the world twice, it seems. And while I'm certain my grandchildren may enjoy your dating PrefX, the company I speak of is Sol-vang. You didn't take the expected path, and we are all better because you led the way."

He turned and faced the group. "I have to admit, never in my life would I have believed I would receive both the Labonne and Seng families as my guests."

He gestured toward the chairs arranged around a long, oblong table. President Batista sat at the head, Philippe and Angélique sat on his left, and Preap and Kasemchai on his right.

President Batista sat up straight and addressed the group. "In my capacity as president of Brazil, I am reminded daily that I am not a great man but that I have a great responsibility. I serve the people of Brazil. And seated at this table, I am Brazil, and on my left"—he raised his left hand in a sweeping motion—"we have the Labonne water family. And on the right"—he raised his right hand—"we have the Seng water family."

He dropped his hands. "It is my duty to protect the interests of the Brazilian people while finding a path that brings prosperity to your two families. Water is the source and necessity of life. It is also mysterious. Do you know there is a moment when water is especially unique? It is called the triple point. That is the moment when temperature and pressure are just right, and water becomes a solid, a liquid, and a vapor simultaneously. It can coexist in equilibrium."

President Batista smiled at everyone in the room. "That is my goal for our meeting—to find equilibrium."

He waved his hand over the table, and two rectangles filled with text floated above the experience sphere at the center of the table. Preap and Philippe looked up at the familiar documents.

"I have reviewed both of your proposals." He clasped one hand with the other and placed his balled hands softly but decisively on the table. If I may borrow a wonderful word of yours, Kasemchai, it is 'serendipitous.' The two proposals are remarkably aligned. With the routes you are proposing, north and south, you estimate that your combined efforts would ship approximately ten billion gallons in the first year."

"That's correct, Mr. President," Philippe said. "The value of your new water can only be achieved when it reaches a population that so desperately needs it."

He gave a subtle nod to Kasemchai, who picked up the conversation.

"That's correct, Mr. Labonne. We are committed to ensuring Brazil will be able to deliver water, and our two families have agreed to a commitment of fifty new ships each by the end of year two."

"I see," President Batista said. He glanced at Alessandra. "I suppose I have a question if you'll entertain me. You have proposed exclusive rights to shipping for the next ten years. That's a lot to ask for."

Kasemchai nodded. "It's true." He raised his hand in a cupping motion, but the fingers were almost straight. It was a carefully choreographed move designed to indicate his desire to welcome conversation. "However, we are committing a tremendous amount of resources for you to be able to ship water, which will bring your country immediate revenue. We're also helping you build the pipeline to Peru, and that is a ten-year commitment on our part to complete the project."

President Batista leaned back in his chair.

"I think you can see," Kasemchai said, "that the ten-year commitment is really twofold. It helps us to recoup our losses while allowing Brazil time to stabilize its economy."

"I appreciate that, Kasemchai. What makes this so complicated for us is we are not committed to the pipeline idea. Our feasibility studies on the pipeline have led us to believe there are far more effective ways for us to focus our efforts."

Kasemchai knew the pipeline was a weak incentive for the Brazilians.

He had told the group that he didn't like tying their ten-year timeline to the pipeline because it didn't give him much leverage if it was rejected.

"Perhaps you could tell us what other projects are of interest to your nation?" asked Kasemchai.

"Of course. The truth is, what we want is a shipping fleet. It is a blessing that we have water today, but we can not allow ourselves to be bound to two customers. The world needs water, and we'd like to bring it to the world." He slowly spun the glass of water in front of him around in circles. "We know all too well the pain of not having enough water."

"I see," Kasemchai said as he rested his hands flat on the table. "So a fleet of your own is what you wish to prioritize?"

Philippe and Preap stared at each other. Preap nodded ever so slightly and looked back to Kasemchai.

"It's not as simple as that," President Batista said. "Of course, we want to embrace your support and make it worthwhile. It will take years, perhaps even a decade, but eventually, we need the ability to build our shipping capability."

Kasemchai's face fell. He wanted to glance at his father to judge his response but didn't dare. Instead, he stared down at his fingers laid out on the table. "Should I understand that with your focus on building a fleet, you would wish to forgo the pipeline north? And that you wish to leverage our support"—Kasemchai looked up at President Batista—"to build your infrastructure and build ships to eventually end the need for our ships?"

President Batista pursed his lips. "A pipeline may be a very low-cost and reliable way to send water north, but there are other, more efficient ways to bring water north that do not require such an investment." He moved his open hand toward his heart. "And I am not implying that we would walk away from our partnership. We very much value your investment in our country and cherish the idea of a partnership. Surely you can see that we want to support an open market?"

"An open market with whom?" asked Kasemchai. "We offered—"

Preap interrupted Kasemchai. "In that case, we should focus on your need to create revenue until a time when you do have the ability to ship water. As you saw in our proposal, our ships will bring water to all the main ports, and the value of that water is dependent on the cost of supplying that route. Some routes are more difficult than others. But, in order to keep things fair, we will establish a base rate for receiving the water in Brazil. This base rate will be the same for the two water families and Brazil, assuming at that time that all three of us will be able to ship your water from that point forward."

President Batista smiled broadly. "I love this spirit of collaboration. There's just one detail I'm a little unclear on. How do we establish a base rate for both of you and Brazil or should I say, all three regions? In your regions, your own ships will move your water. So, the base rate is not a cost for you."

Preap cleared his throat. "You are correct; we do not sell the water to our own ships. However, we face the same transportation costs of delivering water, which includes our substantial investments in setting up the ports. So, the base rate will enable us to determine a base rate for selling the water to other countries and help us establish a fair market rate for all three regions."

"I still don't follow what you mean by a base rate. Our ability to determine a price for our water will be set by you and Philippe."

Preap placed his palms flat on the table and spoke in a straightforward manner. "It's really quite simple. Your country currently has an abundance of water for the first time in almost two decades. We are proposing an opportunity for you to elevate your country out of poverty. In a very short period of time, your country can achieve prosperity and influence never seen before in your country's history. You lack the resources and connections to transport water to our ports located up and down the coast."

President Batista's eyebrows shot up. "What do you mean, *your* ports?"

Philippe grinned. "The world is a complicated place, Mr. President. How do you think Sol-vang was able to build ECs in all the major ports of the world in only a few years? How do you think we were able to ship water so easily around the world? We started as a water company. Our primary business is distribution. And let me be totally clear on this point: we decide which ships can dock in our ports."

Kasemchai looked up at Philippe and then at his father. Very slowly, his face transformed as he realized that even his greatest accomplishments were completely beholden to his father's whims.

President Batista's face stiffened. "Are you implying that all the water can only be delivered by the two water families?"

"It's far more than just delivery," Preap explained. His tone was neither condescending nor inviting for further negotiation. "It's fundamental to our brand. We provide the world a choice, but ultimately, people can only manage between two options. This approach is what's best for everyone."

Kasemchai pivoted his attention to Angélique. She smiled back at him with her wicked smile. She was loving this. He realized that she had known all along. He was not there to lead the negotiation, only to make sure Liv didn't show up. And they hadn't even left that to him.

At that moment, another thought came to him. Liv had said he needed to do what the water families wanted. Sol-vang was simply a tool for their scheming. The problem for them was Liv.

She had been standing beside him this entire time. They built Sol-vang together. Liv was the only one he could trust, and they had—he corrected himself—*he* had removed her from the negotiation.

Kasemchai tapped his AugWatch and pretended to read a message. He stood up and pushed his chair back, startling everyone. He clasped his hands in front of his face and bowed, first to the president and then to the fathers. "I apologize. I have just become aware of an urgent matter I must attend to. Would it be possible for us to reconvene in a few hours?"

President Batista glanced at Alessandra and then back to Kasemchai. "I think that would be a good idea. We still have a lot to discuss, and we have a lot of interesting information to process."

Chapter Sixty-Nine

KASEMCHAI RACED to Liv's hotel. The car had barely pulled to a stop when he threw open the door and sprinted inside. The floor of the lobby had a circular design with couches and tables arranged around concentric circles. Kasemchai didn't slow and weaved around the furniture, calling Liv on his AugWatch as he made his way toward the elevators on the far side of the room.

Liv's avatar appeared and began to run beside him as he reached the halfway point of the room.

"Hey, Kasemchai. I'm headed to the airport shortly. What is it?"

Kasemchai stopped running and tried to sound nonchalant. He was standing uncomfortably close to a woman sitting in a chair. Her son was standing in front of her and stopped complaining to stare at Kasemchai. He looked confused. She looked concerned.

He looked down at the boy, gave him a wink, and started walking briskly again toward the elevators. "Oh, hey. I knew you were going to the airport soon. I just wanted to drop something off for you. A little gift."

"A gift? What on earth for?"

"Oh, it'll make sense when you see it. What room are you in?" He started walking toward the elevators again.

"I guess that's nice, Kasemchai. But like I said, I'm leaving soon. It's really too late."

"That's okay. The messenger is down in the lobby now. Could you give me your number so he can run it up?"

She sighed. "It's 6090."

Kasemchai nonchalantly reached out to the elevator keypad and selected the sixth floor. When the doors opened, Liv's avatar followed him into the elevator, and they started up.

"Aren't you going to ask me how the negotiations are going?" he asked.

"No. No, I'm not. I don't want to know."

When the elevator door opened, Kasemchai ran out. He stopped in front of the numbering plaque and tried to decipher which direction was 6090, then dashed down the hall, Liv's avatar running beside him.

"Ah, well . . ." he panted out. "I guess it's probably for the best."

"Are you still in the negotiations?" she asked. "You sound winded. What's going on, Kasemchai?"

He stopped in front of her door and placed his hands on his knees, catching his breath. "Oh, it's just all the stairs here."

"Look, we should hang up. I told you I couldn't be part of the negotiations."

Kasemchai stared at the door. He placed his fingers on the numbers and traced them slowly. "Look, I wanted to tell you something. When we last spoke, you asked why I never told you how I ended up on the dive boat. And I said we got busy. But it was a good busy, right?"

Liv's avatar looked up, searching for something. "Yeah, it was a good busy."

"Liv, here's to courageous change."

He reached out and knocked on the door three times.

"Courageous change? Wait, hang on, Kasemchai, your messenger is here."

Kasemchai and her avatar stood side by side, staring at the door.

It opened, and Liv stood there.

"What?" She stepped back, her eyes darting back and forth before locking onto his. Her nostrils flared as she looked away, and then she fixed him with a stern glare. "You can't be here, Kasemchai. Angélique was very clear . . ."

"Look, Liv, I . . ." His gaze bounced from the door to down the hall. "Could I come in, please?"

"I told you Angélique will cancel her contract with us if I talk to you about the negotiations. This would destroy Sol-vang."

"I want to talk to you about something more important than Angélique."

Liv shook her head. "Kasemchai, thousands of people rely on our company for their livelihoods. Whatever you want to talk about can't be more important than the people who depend on this company."

"Fine." Kasemchai glanced down the hall again, trying to gather his thoughts. "Do you see me as a partner? What I mean is—" He wrung his hands. "Am I just an investor? Or do you see me as more than that?"

"I don't know why you're asking me this." She sighed loudly. "Kasemchai, I don't know why you put me in charge of the company. Or maybe I do, but the thing is, you followed through. You didn't just make me president. You supported me. You pushed me." She wiped a tear from her eye. "You're the most important person in my life, Kasemchai."

Kasemchai grinned. "That's what I wanted to hear, Liv."

Liv didn't smile back, and her face reddened. "What do you want from me? Why are you asking these stupid questions?"

"I'm about to make the most important decision of my life, and I needed to know where we stand."

Kasemchai glanced at the bed, where, as expected, Liv's bag lay open, and her clothes were unpacked and scattered. He turned to the table by the window and noticed a mug with steam rising slowly. "Did you burn yourself on the tea again?"

She laughed and patted his chest lightly. "Stupidity is doing the same thing twice and expecting a different result."

Chapter Seventy

EVERYONE LOOKED up when Preap entered the room.

"Any word?" asked President Batista.

"None. We'll have to start without him."

Preap walked towards his seat and sat down.

Philippe stood and pushed his chair back slightly. "I suggest we restart the conversation where we left off about the rebuilding of the Brazilian infrastructure. Preap and I have learned through much trial and error the most efficient way to harness the vast quantity of water and where you'll need to put in place water purification plants."

President Batista smiled. "I don't believe that subject has been addressed yet, but let's begin there."

The door swung open at the end of the room, and Kasemchai entered. A sly grin crept across his face, and he strode towards the great table. He left the door open, and Liv entered the room behind him.

Several people in the room gasped.

Kasemchai held a hand aloft. "Ladies, gentlemen. After listening to the course of this conversation, I realized that we were missing someone with much at stake and much to add to this conversation."

Angélique pounded her hand on the table. "This is ridiculous! She cannot be here. We each agreed only two of us would participate in the negotiations."

Kasemchai checked the cuffs of his shirt. "Oh, I think you misunderstood me. She's not here to join the Seng family."

Liv walked around the table and stood next to President Batista, who could barely hide the glee on his face. An aide raced forward and brought a chair for Liv to sit down in. After she was seated, Kasemchai sat down beside his father.

Philippe and Preap stared across the table with steely eyes.

Philippe broke the silence. "Liv, I'm certain the Seng family is very happy to have you join our conversation. But I struggle to see how you can be an impartial participant."

Liv crossed her legs, laid her hands on her lap, and leaned to the left. "That's a very good question, Mr. Labonne. Sol-vang's core mission has always been to reduce carbon pollution worldwide. I assume you're aware of our joint venture with the Brazilian government, Brazil-Sol-vang. We're converting heavy military vehicles and building a fleet of battery-powered tankers."

Philippe pursed his lips tightly before saying, "I am aware of that."

"The ships we're converting to battery make it more efficient to run Aframax-class tankers—"

"Why would you build such small tankers?" Philippe interrupted.

"With a draft of fifteen meters, the Aframax-class vessels can dock in far more ports—ports that neither the Seng family nor the Labonne family currently support."

Philippe clenched and unclenched his jaw. He turned and looked over at President Batista. "I recognize the potential for you to ramp up shipping on the Brazilian coast. As you know, Mr. President, we have also proposed to invest in your infrastructure, but I propose we make an adjustment to our investment. I would like to reduce our investment."

"Reduce it?" asked President Batista.

"Yes, my friend. I would like to divert that commitment. You have many years ahead of you to harness the flow of rain while repairing your country from the damage from the floods. You want to build ships to raise capital, and I can ship water now. We will make a commitment to purchase eighty percent of your ships over the next five years." He directed his attention to Kasemchai. "And Kasemchai, it appears you may be the savviest investor alive—because you will benefit heartily from this deal."

Liv cleared her throat. "Mr. Labonne, that is a very generous offer. And there's no question that Brazil can use this generous commitment at this tumultuous time. But President Batista has made one thing very clear: Brazil doesn't want to sell the vessels."

Philippe's smug smile quickly faded as he said, "You do not speak for Brazil, Liv. This is a very generous offer."

Angélique bared her teeth. "This is not fair."

Kasemchai turned calmly toward Angélique. "What is not fair?"

"You will control everything." Angélique pressed one hand down on the table as if she was trying to crush it. "The Seng family will own all the water in Asia and benefit the most from the Sol-vang Brazil venture. We cannot allow this power grab."

A mischievous smile crept up Kasemchai's face. "Wasn't that your plan, Angélique? To split everything between the two of us?" He met Angélique's stare. "But that is not what's happening."

Kasemchai turned toward Liv and nodded. Then he tapped his AugWatch and reached out into the air in front of him to send a message. Within seconds, everyone's AugWatches began to buzz, and they looked down frantically.

Liv and Kasemchai stood up calmly and walked toward the end of the table opposite President Batista.

An aide ran to remove Liv's chair beside President Batista while two more aides rushed forward with seats for Liv and Kasemchai.

After sitting, Kasemchai poked at the air, and a news article jumped to the experience sphere in the center of the table. "As you can see in the press release you are no doubt reading on your AugWatch, I have stepped away from the Seng family business to redouble my efforts to support Sol-vang. Furthermore, I have divested all shares in the Seng family business."

He stood, placed his palms together in front of his face, and bowed very deeply toward his father.

Preap's face twisted. "What are you doing, Kasemchai?"

Kasemchai bowed to Philippe and then to Angélique. Then, he

turned back to his father. "You said the rain was a gift. What you and Philippe did was hard. The world was in chaos. And every step of the way, people tried to steal the water and destroy your efforts. If you didn't succeed, we all failed."

Angélique shifted in her chair. If she could spit venom from her eyes, she would have.

Kasemchai continued, "You and Philippe cannot be the only two sides, even if you have humanity's best interests in mind. Your solutions always support one inevitable outcome."

Angélique stared hard at him. "And what is that outcome, Kasemchai?"

"Every debate and decision has only one goal: stay in power. And why not? They have made all the right decisions. But unfortunately, no one ever wants to compromise on power."

Kasemchai nodded to Liv. She swiped her hand over her AugWatch and sent a message. Within seconds, everyone's AugWatches buzzed anew. Liv swiped a hand over her AugWatch again and tossed an article to the experience sphere.

"This second news article says—and please forgive me for jumping to conclusions, but I felt our negotiations were leading in this direction —Brazil, the Labonne, and the Seng families are committed to bringing water to the world. Brazil has agreed to give the Labonne and Seng family exclusive rights to ship water to their ports for the next ten years."

Liv regarded President Batista, who now looked as confused as Preap, Philippe, and Angélique. "It goes on to say Brazil and Sol-vang will use a significant part of those funds, with a matching contribution from China, to build a fleet of electric Aframax-class tankers to bring water to hundreds of new ports."

"There's one more part," Kasemchai said. He scrolled down the article floating above the table and stopped to read. "'It is the mission of the three'—and I really love the term they've come up with here, Water Carriers—'the three Water Carriers to collaborate closely on solutions to bring water to more people. And it is believed that the escalation of ships converting to battery power will bring rain to new regions in the future."

Angélique placed both palms on the table, steadied herself, and hissed, "You have no goddamn idea what will happen to the environment. And you've walked away from your family, *our* families, to pursue this absurd idea."

Kasemchai carefully sat back down. "You're right. I don't know. But I know that Sol-vang will support the need for all three regions to ship as much water to the world as possible. And that ship traffic will increase while the cost of shipping decreases." He looked back at his father. "I'm sorry, but it's true. There needs to be more than two sides in every important decision. Hell, the Saudis still have tons of investments. They could revive their economy by creating a battery infrastructure just like Sol-vangs. There's an infinite world of possibilities."

Kasemchai picked up the glass of water in front of him and drank it. It was the sweetest water he had ever tasted.

The room was silent as each party glanced from face to face. A thousand words were considered, but not a single one uttered.

President Batista finally broke the silence. "I, on behalf of Brazil, will support the statements shared in this news article."

The room remained silent.

Kasemchai asked, "Are there any objections to what was laid out in this article?"

Again, silence.

Kasemchai stood and held out his arm to Liv. "I'm famished. Would you like to join me for lunch?"

Liv stood, and together, they walked out the doors.

In the hallway, Liv asked, "So, are we still doing something in April this year? It's become kind of a tradition."

Kasemchai smiled. "Are you familiar with Lent?"

Liv laughed. "You're Buddhist, Kasemchai. I'm not exactly a devout Catholic, but yes, I know what Lent is." She playfully jabbed his ribs with her elbow. "Are you thinking of giving up something for Lent?"

Their footsteps echoed in the empty hall, and a mischievous spark danced in Kasemchai's eyes. "Well, as you said, I'm a Buddhist, so I don't have to give up anything for Lent. But I know of a really good party that happens before Lent."

Chapter Seventy-One

THE SOL-VANG RESKILLING Center was just outside the Port of Recife. The building had to be completely rebuilt. Many of Recife's buildings had been abandoned during the drought, but now it looked like the entire city was under construction.

Makoa and Kasemchai sat on tall stools at the front of a class and were peppered with questions from the students—many were workers from the port, but not all. The reskilling program was attracting a lot of attention from people looking for a solid career in the new world.

They took a last question, and Kasemchai thanked the people in the room. All smiles, he and Makoa backed out and walked down the hall to the bridge inside the atrium.

Makoa stopped to look down at the students ambling in and out. "I love seeing all these people down there, Kasemchai."

Kasemchai didn't look up. "Why is that?"

"It's hope. They have hope for the future. There are so many things going wrong right now. So many of them have lost someone up north or know someone who's been displaced. As usual, the world is coming down on their shoulders, but they actually have hope."

"Where are you headed to next?" asked Kasemchai.

"We're opening reskilling centers near the Port of Aratu, Port Santos near São Paulo, and the Port of Pelotas. Then we head to Port of Montevideo in Uruguay and finally down to Argentina."

Kasemchai smiled at Makoa. "Wow, that's a lot of travel."

"To tell you the truth, before we came to Brazil, I'd never been to a foreign country."

"What?"

"Not everyone gets to fly around the world like you do."

Deep in thought, Kasemchai finally asked, "How do the kids like school?"

"I was a bit worried about them going to an international school in São Paulo, but this is the best school they've ever been to in their life." Makoa's smile vanished, and his brow knitted. "When I told you about the infinite universe of possibilities, this wasn't one I ever expected. I know I've said this before, but . . ."

"It's okay, Makoa."

"No, man. You asked me to be your conscience, and your conscience says you really did something amazing—for a lot of people. And I know your dad's proud of you."

Kasemchai chuckled. "Yeah, he's still not talking to me, but Mom says he's coming around. To tell you the truth, I don't think he's so mad that I left the business but that I outsmarted him."

"He has to respect that, Kasemchai."

Kasemchai smiled. It was a serene smile, something he did a lot these days.

"So what's next for you?" Makoa asked.

He straightened, slapped his hand on the bridge railing, and said, "Well, it's April again. And that means I have plans with Liv."

Chapter Seventy-Two

KASEMCHAI'S MIRROR-GREEN Maybach EQL Cabriolet flew along the mountain road. Six meters in length, the sloping rear evoked the image of a beautifully lacquered Chris-Craft Runabout. It trailed a crackling resonance, like electricity arcing through the air.

Liv gazed out her window at Rio de Janeiro. The rising ocean had crept into the city, and many of the buildings closest to it had tumbled over. But there were lights everywhere, and the city was alive.

He slowed down and turned into the curved driveway of the hotel —brilliant white with chiseled rectangular lines of columns and windows—elegant yet a little bit naughty.

Liv leaned over to Kasemchai. "Are we really here for Carnival?"

Kasemchai tilted his head toward her and grinned. "Hey, I know how much you like your Easter. 'Carnival' literally means 'farewell to meat.' If you're going to fast for Lent, why not start here?"

The car came to a stop, and valets dressed in white opened the door.

A soft gasp escaped her lips as Liv took in the enormous white hotel. "I can't wait to see my room."

"I couldn't make up my mind." He flourished his hand in the direction of the upper floors. "So, I booked three of the penthouse suites. You can see them all and take your pick."

"I thought you always booked the entire hotel, Mr. Seng."

Kasemchai winked. "It's not a party without people."

The concierge stepped forward. His suit was all white, with shiny buttons working up in a V before the broad lapel. This contrasted sharply with his pink tie and bald head, which reflected the lights from above.

"Welcome, Mr. Seng. And Ms. Anselm, we, of course, know you well."

"Me?" Liv was not the woman she had been when she met Kasemchai, but when Kasemchai entered a room, all eyes were on him. Being in his shadow, she felt, made her more effective. "You do? How?"

"You are very modest, Ms. Anselm. The Brazil-Sol-Vang company is very hot news these days, and you are its leader." He bowed and gestured toward the hotel. "But I'm sure you're not here for work. Please follow me; I'll show you to your suite."

THURSDAY WAS the first night of Carnival. Kasemchai stood outside Liv's door wearing a black tuxedo and an alabaster feathered mask with gold filigree and jewels. The steady beat of the drums and horns made the floor vibrate under his feet.

The door opened, and Liv stepped out in a floor-length dress that shimmered with every movement. Beneath the sheer fabric, he could make out the two strands of chain that wrapped around her waist. Her eyes sparkled behind her mask, and a broad grin spread across his face when he realized she was staring at him.

He reached out to take her hand. "May I be your escort to the party tonight?"

She placed his hand in his. "You may."

Kasemchai's eyes darted behind the hard shell of his mask. "So, Liv. What will you give up for Lent?"

Her lips curled up in an impish grin. "Probably drinking."

Glossary

AI Collaborators: In a world where AI can learn and improve on its own, engineers transition from being the sole creators to collaborators. Instead of managing all technical tasks themselves, they focus on guiding AI's development and ensuring its ethical use—their role shifts to steering AI's potential toward human-centered goals.

AISL: Artificial Intelligence Service Layer. In the future, a series of services will be required for AI to provide a robust experience for people, especially when it comes to private information, such as financial or health information, placing orders at a store or restaurant, or enabling personalized services, such as dating. The AISL provides the UI/UX based on user preferences and behaviors. People interact with the AISL, which can use many different AI models from the AI Personalization Layer. Below the AIPL are the compliance and regulation layer (which may be religious, government, or ideology), the security and privacy layer, the data integration layer, and finally, the vendor and data sources.

Aframax Vessel: A type of oil tanker that can carry up to six hundred thousand barrels of oil, or twenty-five million, two hundred thousand gallons of water. Due to its shorter length than the Panamax class vessel, it can serve most ports in the world. An Aframax vessel can be up to eight hundred twenty feet in length, one hundred thirty-eight feet across the beam, and have a draft of up to forty-nine feet.

Apsara Dance: The term "Apsara," meaning "celestial nymph" in Sanskrit, symbolizes the divine connection between heaven and earth, with over fifteen hundred unique carvings at Angkor Wat inspiring the Royal Ballet of Cambodia. The dancers, adorned in golden headdresses and silk dresses, perform intricate hand gestures and flowing footwork in "the dance of the celestial nymphs."

AugWatch: The colloquial term used after the first augmented reality watch was released. The disc on the top can be removed and placed on a table to place the view directly in front of the user or to share information with other people.

Byte: Slang for ass. Originated from people who shared too much data about themselves on social media. It became a cheeky way to refer to someone who was being an ass.

Bucha: Slang for Kombucha. It's an alcoholic beverage, generally five percent alcohol, but it can pack more punch with boosters.

Butazeri: A type of meat substitute similar in consistency and flavor to Spam. It's a mashup of the Japanese words "buta," which means "pork," and "zeri," which means "jelly."

Directographer: Using camera drones and light drones, directors can control all aspects of creating a Virtumersion. The name combines the roles of the film director, cinematographer, camera operator, lighting team, and sound team.

Energy Center: A location created by the Sol-vang company to store and repair batteries before they are recharged and returned to service ships. They need to have a reliable source of electricity and can be located a short distance from the port to keep costs down.

Experience Sphere: An alternative to the AugWatch that allows many people to see or experience anything digital from a communication to Virtumersion.

Fish Milk: A milk substitute made by deboning and grinding fish into a powder, then mixing it with water, emulsifiers, thickeners, and flavorings. It's high in protein and omega-3 fatty acids.

IE Gel: This snack is specifically designed to provide a quick energy boost for athletes and individuals on the go. It is rich in protein, vitamins, and minerals, and also contains fast-acting carbohydrates such as sugar or glucose syrup, along with caffeine and electrolytes.

Luddites: The Luddites were nineteenth-century English workers who protested industrial machinery, fearing it would replace skilled jobs and threaten their livelihoods. While their concerns about job displacement were valid, they underestimated how technological progress could improve living standards and create new opportunities over time.

OPEC: An acronym for the Organization of the Petroleum Exporting Countries. An alliance of mostly Middle East and South American countries that determines the price of oil globally by increasing or decreasing oil production.

Panamax Vessel: Ships that must pass through the Panama Canal lock chambers are regulated by the Panama Canal Authority. A Panamax vessel can be up to nine hundred fifty feet in length, one hundred and six feet across the beam, and have a draft of thirty-nine-point-five feet.

Photo Bio-reactor: A tank of liquid filled with microalgae. They were common in dense cities that no longer had trees to help consume CO_2 and oxygenate the air.

PrefX: An amalgamation of preferences with experience-centric design. It allows people to share how they view and interact with data. People can share their PrefXs on an AISL. PrefXs can also be monetized to allow anyone to make money from a clever AI-designed interface and experience.

Seaweed Sachet: Containers for holding liquids like water or kombucha. The odorless and tasteless containers are soft like a plastic bag but made from seaweed. They sink in seawater and are biodegradable after two years. They are also edible and high in fiber and essential vitamins.

VidSheet: A paper-thin sheet that allows you to view a document or video.

Virtumersion: A digital experience format. Instead of a two-dimensional experience, people participate in a fully immersive and sensory-rich experience from their AugWatch or Experience Sphere. Virtumersion allows people to watch in movie-screen format or step directly into the middle of an augmented reality or virtual reality view. Using AI, all people involved with production are removed from view, users can talk and manipulate the environment, and viewers can create new content based on the guidance or story background provided by the directographer.

Characters

- **Aksel** (AHK-sul) Rotterdam, The Netherlands - Liv's colleague who works on government contracts.
- **Alessandra** (ah-leh-SAHN-drah) Brasília, Brazil - Advisor to president Batista.
- **Amporn** (am-PORN) Preap's AI companion.
- **Anela Malo** (ah-NEH-lah MAH-loh) New Jersey, USA - Makoa's wife. Mother to Kailani and Apikalia.
- **Angélique Labonne** (ahn-ZAY-leek lah-BON) Anyama, Ivory Coast - Daughter of Philippe and Sophia.
- **Annop** (ah-NOP) Sihanoukville, Cambodia - Kasemchai's friend from the Aquadrome.
- **Apikalia Malo** (ah-pee-KAH-lee-ah MAH-loh) New Jersey, USA - Makoa and Anela's younger daughter.
- **Arlo** (AHR-loh) Rotterdam, The Netherlands - Executive Chef of the secret restaurant.
- **Aroon** (ah-ROON) Bangkok, Thailand - Young boy at the Songkran (Thai New Year) festival who helped Mari, the young Thai girl.
- **Bai** (bye) Georgia, USA - Student at the University of Georgia.

- **Ben** (BEN) New Jersey, USA - Deck officer of the bunker barge who first greeted Makoa.
- **Bob Hamilton** (BOB HAM-il-ton) USA - President of the United States of America.
- **Brechtje** (BREKH-tyuh) Rotterdam, The Netherlands - Base jumping friend of Liv.
- **Captain** (KAP-tin) Ishigaki Jima, Japan. Captain of the dive boat.
- **Carina** (kah-REE-nah) Brasília, Brazil - President Batista's AI companion.
- **Champei Seng** (cham-PAY Sung) Sihanoukville, Cambodia - Kasemchai's mother.
- **Chantal** (shahn-TAL) Rotterdam, Netherlands - Liv's colleague.
- **Chantana** (chan-TAH-nah) Sihanoukville, Cambodia - Leader from Personaspace who pitched the AISL business to Kasemchai.
- **Chantavy** (chahn-TAH-vee) Sihanoukville, Cambodia - Kasemchai's AI companion.
- **Darius** (DAH-ree-us) Ghawar Field, Saudi Arabia - Egyptian roughneck on the oil crew.
- **Duncan** (DUN-kan) Maine, USA - Pro-desalination senator ruined by Philippe and Preap.
- **Emma** (EM-mah) Rotterdam, The Netherlands - Attorney at Sol-Vang.
- **Emmet** (EM-it) Newark, USA - Owner of the Newark Winery.
- **Evelyn** (EV-uh-lin) Georgia, USA - 911 AI operator.
- **Faya Camara** (FAH-yah kah-MAH-rah) Anyama, Ivory Coast - Angélique's husband.
- **Finn** (FIN) Rotterdam, The Netherlands. Liv's colleague.
- **Fran** (FRAN) New Jersey, USA - Office manager at the tankerman office.
- **Fulin** (foo-LEEN) Sihanoukville, Cambodia - The big man from Personaspace who pitched the AISL business to Kasemchai.

- **Harper** (HAR-per) Georgia, USA - Student at the University of Georgia.
- **Haruka** (HAH-roo-kah) Ishigaki Jima, Japan - First dive guide who worked with the Captain.
- **Helena** (heh-LAY-nah) Rotterdam, Netherlands - Liv's colleague.
- **Hendrik** (HEN-drik) Rotterdam, Netherlands - Liv's assistant at Sol-vang.
- **Jacob** (JAY-kub) Ghawar Field, Saudi Arabia - Philippine driller on the oil crew.
- **Jia** (JAH) Dalian, China - The daughter of the security manager, Xiaomeng.
- **Johannes Nielsen** (yo-HAH-nes NEEL-sun) Rotterdam, Netherlands - Liv's colleague who pitched government and military contracts.
- **Kailani Malo** (Kah-ee-LAH-nee MAH-loh) New Jersey, USA - Makoa and Anela's older daughter.
- **Karim** (kah-REEM) Ghawar Field, Saudi Arabia - Egyptian roughneck on the oil crew.
- **Kasemchai Seng** (ka-sem-CHAI Sung) Sihanoukville, Cambodia and New York, USA - Son of Preap and Champei. Sato calls him Kasem-chan.
- **Kato** (KAH-toh) Ishigaki Jima, Japan - Kasemchai's attendant at the hotel.
- **Keone** (keh-OH-neh) New Jersey, USA - Makoa's brother.
- **Koffi** (KOH-fee) Anyama, Ivory Coast - Member of the video crew.
- **Kosal** (koh-SAL) Sihanoukville, Cambodia - Preap's head of staff.
- **Liv Anselm** (LEEF AHN-selm) Rotterdam, The Netherlands - Engineer at Sol-vang who dove with Kasemchai.
- **Lucas** (LOO-kahs) Brasília, Brazil - Secretary of Defense Products - Proposed Brazil / Sol-vang partnership.
- **Luuk Visser** (LUKE FIS-ser) Rotterdam, The Netherlands - Founder of Sol-vang.

- **Makoa Malo** (mah-KOH-ah MAH-loh) New Jersey, USA - Anela husband. Father to Kailani and Apikalia.
- **Mali** (MAH-lee) Bangkok, Thailand - Young girl at the Songkran (Thai New Year) festival who was helped by the boy, Aroon.
- **Nakamura** (NAH-kah-MOO-rah) Ishigaki, Jima, Japan. Manager of the ryokan where Liv stays.
- **Naomi** (nay-OH-mee) Georgia, USA - Student at the University of Georgia.
- **Nary** (nah-REE) Sihanoukville, Cambodia - Head of transportation for Preap.
- **Nee** (NEE) Bangkok, Thailand - Young girl at the Songkran (Thai New Year) festival who is friends with Mali.
- **Nisay** (nee-SAY) Sihanoukville, Cambodia - Preap's driver who took Kasemchai and Angélique downtown.
- **Noa** (NOH-ah) Rotterdam, The Netherlands - Liv's AI companion.
- **Omar** (OH-mar) Ghawar Field, Saudi Arabia - Egyptian assistant driller on the oil crew.
- **Peter** (PEE-ter) New Jersey, USA - Owner of the bunkering company where Makoa works as a tankerman.
- **Philippe Labonne** (fee-LEEP lah-BON) Anyama, Ivory Coast - Head of the African water family. Angélique's father.
- **Preap Seng** (pree-AP Sung) Head of the Cambodian water family. Kasemchai's father.
- **President Ndlovu** (Prez-i-dent Ni-lo-vu) President of Anglola.
- **Rangi** (RAH-ngi) Newark, USA - Makoa's friend who worked at the Côte de l'Est Vinery.
- **Renato Batista** (heh-NAH-toh bah-TEES-tah) Brasília, Brazil - President of Brazil.
- **Ryan** (RYE-un) Georgia, US - Lizzie's AI companion.
- **Samnang** (SAM-nang) Sihanoukville, Cambodia - The tall man from Personaspace who pitched the AISL business to Kasemchai.

- **Sarin** (sah-REEN) Krong Battambang, Cambodia, Cambodia - Auntie celebrating Choul Chnam Thmey (Cambodian New Year) with her family.
- **Sato** (SAH-toh) Kyoto, Japan. Kasemchai's business mentor.
- **Songkran** (song-KAHN) Bangkok, Thailand - Young boy at the Songkran (Thai New Year) festival who knocked over Mali, the young Thai girl.
- **Sophia Labonne** (fee-LEEP so-FEE-ah lah-BON) Anyama, Ivory Coast - Angélique's mother.
- **Sophie** (so-FEE) Rotterdam, The Netherlands - Liv's colleague.
- **Tanaka** (tah-NAH-kah) Ishigaki, Jima, Japan. Virtuoso distiller at the awamori distillery where Kasemchai first meets Liv.
- **Tiffany Day** (TIF-uh-nee DAY) New Jersey, USA - Republican senator frustrated that the tankerman were standing in the way of progress.
- **Tiken** (TEE-ken) Anyama, Ivory Coast - Directographer filming Look Up to God for the Labonne family.
- **U'ilani** (oo-ee-LAH-nee) New Jersey, USA - Makoa's older sister.
- **Xiaomeng** (SHYOW-mung) Dalian, China - Security manager of the location in China where all of Preap's engines are stored.
- **YanYan** (YEN-yen) Dalian, China - Inspector of the location in China where all of Preap's engines are stored.
- **Yara** (YAH-rah) TK, Brazil - The little girl who sees rain on Easter Day.
- **Young-Soon** (YUNG-soon) New Jersey, USA - Anela's AI companion.
- **Yuki** (YOO-kee) Ishigaki Jima, Japan - Second dive guide who worked with the Captain.
- **Yuting** (YOO-ting) Dalian, China - The daughter of the inspector, YanYan.

Acknowledgments

Thank you for allowing these pages to be a part of your world, even for a little while. It means a great deal to me that you took this journey, met these characters, and devoted your time and thoughts to them. I hope their story has added something special to your own.

I would also like to thank those who helped me bring this story to fruition. A big thank you to my friends, especially Ed Castro, who helped me with the roughest idea to the final manuscript to my website. Shiv Gajula frequently brainstormed with me about the future of technology and AI. And Sean Murphy, who helped me find the missing scenes and convinced me to return to my original ending. My copyeditor, Clara Abigail, was instrumental in refining the structure and encouraging me to flesh out the narrative with essential details. My proofreader, Amy Lisane, gave this story a brilliant shine.

To my friends and colleagues who entertained my many random questions, now you finally understand my mischievous purpose.

About the Author

Gregory Greunke is an award-winning author who explores the space where nature, technology, and the human spirit meet. Born in California, in an area best known for orchards, he's traveled to 32 countries, lived in Australia and Asia, and returned to where his journey began - now known as Silicon Valley.

The Water Carriers is the first book in the *Rain* series. Greg is also developing a YA dystopian thriller set 100 years in the future. He works for a large technology company and carves out time for writing early in the morning and on weekends. He loves spending time outdoors with his family. But every day, he's happiest near water—on it, in it, or under it.

I'd be honored if you shared your thoughts in a review with Goodreads or your favorite book seller.

Subscribe to my newsletter
www.gregorygreunke.com

BIN TRAVERLER FORM

Cut By: Cesar Flores **Qty** 40 **Date** 10/7/26

Scanned By:________ **Qty**____ **Date**____

Scanned Batch ID's

Notes / Exceptions